THE GENERATIVE EDGE

Future Proof Your Career and Earnings

B. Jackson

Printed in the country of purchase.

For more information, contact:

thegenerative.edge@gmail.com

Book design by T.H

Cover design by T.H

Illustrations by M Kerney Art

ISBN - Paperback: 978-1-7637439-0-8

ISBN - Hardcover: 978-1-7637439-1-5

First Edition: October 2024

CONTENTS

INTRODUCTION

If you follow the news, you know that AI is the topic of the moment. Everyone is talking about the current trends in AI technology and speculating about what an AI-driven future will look like. The transformative power of AI is unquestionable; it will change how we operate in both our personal and professional lives, automating the kinds of tasks that take up excessive amounts of our time and freeing us up to spend more time doing the things we love.

Nowhere is the rise of AI being felt more acutely than in the professional sector. Employees and companies across industries are taking notice of the AI revolution and coming to the same conclusions: AI is here to stay, and with its transformative power, it's time to start learning how to integrate it into our workplaces. Employees now face a crossroads: adapt and harness the power of AI or risk being outcompeted by those who do. The rapid integration of AI into various sectors has created a divide between those who skillfully navigate this new landscape and those who remain on the fringes. In this era of rapid change, the message is clear: to stay relevant and excel in your career, integrating AI into your daily work is not just an advantage; it's a necessity.

For those who choose to embrace AI, there are enormous potential rewards. Imagine a future where you have successfully integrated AI into your work process. You will be able to sleep soundly at night without worrying about menial tasks or feeling like you're working tirelessly without making any professional or financial progress. Thanks to your collaboration with AI, you will be confident and unafraid, knowing you are making informed decisions. Automating your workload will give you more free time to spend with your loved ones, and you will be able to stay informed on current trends and be prepared to adopt new transformative technologies as they arise. You will be more productive at work, producing higher quality work in less time, which will be noticed by your supervisors and lead to promotions and raises. You will be an AI-savvy employee, and you will be a leader in the AI revolution.

How does that vision of the future sound? If it sounds pretty awesome, it's time to begin your AI-readiness journey. The Generative Edge is here to be your guide, your gateway to transforming yourself from an AI novice into a savvy user. In The Generative Edge, we will take a structured approach to your AI training. We'll begin with the basics, an introduction to the current AI landscape and a brief overview of what AI is, where it comes from, and how it works. After that, we'll dive right in. We will discuss how to improve productivity with AI, overcome common integration challenges, and review AI-driven development case studies. After that, we'll get beyond the basics, looking at ChatGPT and other natural language models as well as specific strategies for integrating AI into data-driven decision-making, financial oversight, and communication. Finally, we'll zoom back out and discuss ethical AI use, prepare for the AI revolution by developing an AI-ready mindset, and examine the future of AI.

Along the way, we will build your 'AI Toolkit.' You will enhance it with skills and knowledge to kick-start your journey into the world of AI. This knowledge will come in two parts: theoretical and practical. The theoretical knowledge will help you create your AI integration vision; you will learn what parts of your job could be automated with AI, how to use AI to make informed decisions, how to optimise team assembly with AI, and much more. Practical additions to your toolkit will include specific software that you can use in an array of AI integration applications, as well as a list of 100 ChatGPT prompts to help you use the chatbot more efficiently and an AI Integration Starter Assessment that will give you a structured system by which to approach AI integration.

The Generative Edge is your guide to transformation. By the end of this book, you will be equipped with the tools to envision your future self and turn that vision into reality. With AI integration, a reinvented version of yourself is within reach.

Let's go find your edge!

1
EMBRACING THE AI REVOLUTION IN THE WORKPLACE

For a long time, the notion of artificial intelligence as an aide to workplace efficiency was a dream, the territory of wishful thinking from overworked employees who would say, "One day, I'll get a computer to do all the menial tasks I don't want to do." However, the early 2010s saw a breakthrough in AI technology that transformed the field from a niche service useful in a few computation-heavy fields to a general-use service capable of growing and responding to society's evolving needs.[1] This breakthrough is called "deep learning," it represents a machine's ability to gather multiple layers of information from a single user input.[3] For example, when an image is put into a deep learning machine, the machine may use a low layer of analysis to process edges. In contrast, a high layer of analysis can conduct facial recognition. Deep learning has allowed for training in which a machine learns to produce new information by taking in and analysing user inputs.

A machine's ability to produce high-quality content is revolutionary, and in the past few years, workplaces have started to take notice. In a 2023 survey of over 1,300 U.S.-based companies, 90% said they aim to benefit from AI-powered solutions by 2028.1 These companies suggested that AI would allow them to automate repetitive tasks, facilitate employee learning, and encourage innovation.1 Additionally, 73% of surveyed companies planned to prioritise hiring individuals who demonstrate savvy AI use.1 Since so many companies want to hire people who know how to use AI to their advantage, building your AI skill set will offer you a tremendous advantage when competing for new jobs or demonstrating your value to a current employer.

AI will inevitably reshape the workplace, and it will likely have transformative powers that we cannot begin to comprehend at present. Even so, there are several theories on how AI will create the workplace of today, tomorrow, and the near future. These include:

- Reversing roles. We will likely reach a point where AI does most of our work, and many human positions will be geared towards simply supporting that work.[2]
- Shifting means of production. We will reach a point where one AI machine can do the work that requires several human employees. This will require employees to re-evaluate how they build their skill sets to complement AI without being rendered redundant.[2]
- Disrupting linear career paths. With refined AI, personalised learning experiences and AI-powered peer-to-peer learning will prepare employees to move into new positions they might have needed to prepare for under traditional learning structures.[2]

The common thread connecting these developments is adaptability. For you as an employee, learning how to identify and implement opportunities to complement the AI revolution will help you remain competitive in a rapidly evolving workplace.

AI in Action: Exploring It's Diverse Uses Across Various Industries

Although AI may appear most useful in the tech sector, its implementation spans various industries, including retail, healthcare, travel, education, and creative fields like fashion. Below are some examples of how AI has become a revolutionising force within these and other industries.

- **Retail:** AI can personalise the shopping experience by scanning customer behaviours and offering customised recommendations.[6] It can also create dynamic pricing environments, increasing the price of a product when it is in high demand and making sales when there is excess supply.
- **Healthcare:** AI is being applied to medical imaging,

analysing CT scans and X-rays to aid in diagnosing conditions like pneumonia, tuberculosis, and cancer.[6] It can also create a personalised treatment programme.[6] For example, a speech-language pathologist may use AI to generate practice exercises for patients recovering from a stroke to rebuild their speaking ability.[6] An oncologist may call on AI to suggest a chemotherapy regimen based on a patient's genetic profile.[6]

- **Travel:** AI can generate a personalised travel itinerary based on a person's interests and budget.[6] Additionally, chatbots on a booking website can guide a traveller through booking flights, hotels, or train tickets, including answering questions about pricing or cancellation policies.[6]

- **Education:** AI can analyse a student's needs and create a personalised learning experience that reflects those needs.[6] It can also be used as an automated grader for both quantitative (e.g., multiple-choice) and qualitative (e.g., essay) responses, thereby freeing up educators to focus on delivering instruction.[6]

- **Fashion:** AI can pull data from social media and fashion blogs to gauge industry trends, which can help fashion designers respond to consumer interests.[6] It can also help with the design process, enabling designers to get suggestions on patterns, materials, and colours.[6]

- **Food technology:** AI can predict supply and demand and manage food inventories to help reduce food waste and get food into consumers' hands more efficiently.[6] Restaurants can also use AI to build personalised menus based on individual preferences, dietary restrictions, and feedback from previous similar customers.[6]

- **Financial services:** AI can protect investors and individual bank accounts through automated fraud detection.[6] For example, an AI bot might analyse a consumer's credit card purchase history to flag an unusually expensive purchase as potentially fraudulent.[6] AI can also participate in customer service by answering account information inquiries and offering personalised financial advice.[6]

Even though AI is functioning differently across various industries, two common themes have started to emerge: efficiency and personalisation. In all these fields, AI is automating the work that would previously have been done by humans, thereby freeing up employees to engage in parts of their jobs that are either more fulfilling, cognitively demanding, or not yet replicable by AI. AI also creates individualised, targeted experiences with wide-ranging implications, from convincing a consumer to spend money to saving a person's life.

Why Understanding and Integrating AI is Essential in Your Career

AI is not going anywhere; in fact, it is already changing the fabric of work as we know it and is poised to have an even more dramatic impact in the near future. Therefore, as an employee in an AI-centric world, you can position yourself to be valued and productive for the rest of your career if you learn to embrace AI and incorporate it into your workflow.

One of the best things you can do right now is start thinking about your field and how to incorporate AI into your work. You may work a job where you do the same kinds of analysis daily. Can you build a module that will allow you to punch in some information and let the module run that analysis for you? Or perhaps you work as an educator. Could you experiment with having a chatbot provide feedback on your students' essays?

Learning how to employ AI can help you boost your productivity and give you the time to engage in the parts of your work that you enjoy the most. Here are some general examples of advantages that AI work integration provides:

- **Administrative tasks.** All the basic stuff you must do to stay organised, like scheduling meetings, sending emails, and organising your calendar, can be completed with AI.

Familiar tools like Microsoft Word, Microsoft Excel, and Gmail are starting to incorporate AI features into their software offerings.[3] Take some time to explore these and other AI capacities to see if you can begin to have AI automate certain parts of your daily workflow.

- **Data analysis.** Have you ever spent hours pouring through data to find patterns or develop predictive models? These days, AI can do both of these things and other quantitative analyses. There are AI bots specifically designed to build predictive models based on inputs from historical data and bots designed to analyse more qualitative content like essays.[3]
- **Research.** Even familiar AI interfaces like ChatGPT can comb the web to locate research on a topic. Some of these sources are becoming faster and more accurate than search engines like Google. You can ask the bot to deliver a specific number of resources on a given topic and to include or exclude particular sources based on perceived credibility. For example, you could ask ChatGPT to provide you with five papers on using AI in the education sector but specify that it should exclude Wikipedia.
- **Learning.** We touched on this in our discussion on AI in the education sector. Still, even if you're not involved in a formal degree program or a continuing education course for your job, you can use AI to explain things you don't understand or to learn a new skill that could help you compete for a more lucrative position.

The Cost of Ignoring AI

With such a strong base of evidence to suggest that employees who embrace AI are poised to succeed in their future career endeavours, it follows that there will be a cost to those who choose to ignore AI. For instance, some sceptics claim that AI should be actively constrained because it will result in job displacement. Since 2010, concerns about AI's pending impact on the economy and the future of work have risen. In 2020, an article in Time reported that "AI job automation has already replaced around 400,000 factory jobs in the U.S. from 1990 to 2007, with another 2 million on the way."[5]

The Time article isn't factually incorrect. Some industries have experienced job losses, and there are strong odds that more job losses will come as AI continues to improve at doing the work that currently keeps humans employed. But this doesn't mean that our jobs are threatened. While it's true that some jobs will likely become redundant thanks to growing AI capacity, employees should not be concerned as long as they find ways to adjust their skill set to complement AI and use the work AI can do to increase productivity in areas where AI is currently less useful.

Others might argue that because AI is still lacking in some areas (like interpreting nuance in certain sensitive situations), it is worth ignoring altogether. However, because AI is capable of deep learning, it will continue to improve until the day arrives when it can be effective in these situations. Furthermore, as AI continues to become ubiquitous in the workplace, workers who choose not to use it may be seen as stagnant or unwilling to change, traits in a world where dynamism and an eagerness to learn are more prized than ever. Therefore, employees must get on the proverbial AI train now while it is still in a somewhat nascent stage. Otherwise, they might get left behind: *"I think workers that don't work with AI are going to find their skills [become] obsolete quite rapidly. So, therefore, it's imperative to work with AI to stay employed, stay productive, and have up-to-date skills."*[4] Carl Benedikt Frey, future of work director at Oxford University.

There is precedent for such a technological revolution and for those who chose not to embrace it. In 1989 Tim Berners-Lee, a scientist at CERN, developed a system by which information could be linked and accessed from a network of devices, not just the ones on which they were stored. When his project was adapted for public use in 1991, it became known as the World Wide Web. Of course, we all know and love the World Wide Web, and we know what a massive shift it caused in the workplace, as it became more accessible than ever to access information and acquire knowledge. These days, it is almost unthinkable to produce work without using the Web, but in its early years, some dismissed the Web as a fad, a file-sharing structure with limited mass appeal that would peter out in a few years. Artificial Intelligence is poised to be the next World Wide Web, the next great technological revolution that changes how we acquire knowledge. Those employees who learn how to use AI as a vehicle to turn knowledge into action will prosper, and those who choose to ignore its potential will almost certainly get left behind.

Summary of Chapter Learnings

Artificial Intelligence is here to stay. It is transforming the way we work, and as it evolves, we will find new ways to deploy it in our continued pursuit of efficiency, personalisation, and liberation from the mundane tasks that so often make up our workday. With the advent of deep learning, AI has gained the ability to take in and analyse a range of information, get feedback on its analysis, and learn from that feedback to produce better results in the future. AI's ability to learn holds unfathomable potential, particularly within the workplace. Far from the domain of tech and other data-heavy industries, AI almost certainly has utility for you as an employee, whether you're analysing X-rays, grading papers, or deciding how to adapt your products to the latest consumer trends. Efficiency and personalisation are two commonalities that echo across AI's deployment. Employers and employees alike are learning how to use AI to speed up, assist with, or completely take over specific tasks, often with more precision than humans, and they are using AI to create individualised experiences, which are likely to be the future of everything from advertising to cancer treatment.

Suppose you're interested in using AI but don't know where to start. In that case, the best introductory implementation of AI is to automate tasks that currently take up a large portion of the day, like writing emails or organising a calendar. However, as AI continues to evolve, you will find the most success in learning to adapt AI to your specific job description, finding ways to automate more parts of your job so you can focus on the parts of your job that you find the most fulfilling. Ethan Mollick, an associate professor at the University of Pennsylvania, says, "What I keep emphasising to people is to just start using this... As workers get increasingly fluent, they can find themselves ahead of the curve and at a distinct advantage in the workplace.

Lean into the AI revolution, become an innovator as you employ it in your daily workflow, and stay adaptable, and you will make yourself an indispensable employee.

2

AI 101: FROM NOVICE TO KNOWLEDGEABLE

What Exactly is Artificial Intelligence?

At its most basic, artificial intelligence is defined as the intelligence of a machine, as opposed to human intelligence.[7] Artificial intelligence is a massive field of study, and it is today an umbrella term under which several subspecialties exist, each of which works towards different goals. For example, some AI developers are interested in teaching machines how to solve problems, while others are working toward teaching machines how to make decisions based on informational inputs. Still, others are teaching machines how to learn or improve their performance on a given task.

A Brief History of Artificial Intelligence

Alan Turing was one of the first scientists to work in machine learning. He developed a theory positing that a machine could use simple symbols like 0 and 1 to conduct mathematical and formal reasoning.[8] This theory started other scientists down the path of attempting to build an "electronic brain."[8]

Turing was thinking about machine intelligence as early as 1941, but AI did not become a mainstream research topic until the late 1990s and early 2000s. During this time, researchers started focusing on using AI to solve specific problems in fields like statistics and economics rather than the broad, versatile appeal for which we have come to know AI today. However, the advent of deep learning (briefly covered in Chapter 1) had the greatest impact on the contemporary AI revolution. Deep learning uses neural networks (which we will learn more about shortly) to automate a learning process after taking in a given source.[8] For example, a machine may be fed a thousand images of dogs to teach it about a dog's basic physical characteristics. After this process, the machine should be able to look at any picture and recognise that there is a dog in the picture. For further learning, the machine could be fed images of different dog breeds to teach it to identify whether a dog is, for example, a German shepherd or a Yorkshire terrier.

Automatic speech recognition (interpreting and responding to human speech) was the first successful implementation of deep learning for public consumption, but several other applications have become commonplace, including image recognition, customer relationship management, medical image analysis, and image reconstruction.[8]

Since 2022, chatbots have taken centre stage in the public discourse surrounding artificial intelligence. Chatbots are machines designed to mimic human interaction with text or voice inputs, essentially creating a conversation between the human and the machine.[10] The most famous of these is ChatGPT. Launched in late 2022, ChatGPT quickly became the fastest-growing software in history, reaching over 100 million users only two months after its debut.[10] The software was noted for being able to generate detailed, human-like text on a variety of subjects; upon its launch, New York Times writer Kevin Roose called it "the best artificial intelligence chatbot ever released to the public."[10] Today, ChatGPT is credited with beginning a new AI boom, which shows no signs of slowing down.

Neural Networks Explained

Think of neural networks as akin to a brain in the machine learning process. A computer's neural network mimics the neurons in our brains by using artificial neurons.[8] Much like the neural network in our brains, an artificial neural network takes in information, learns from that information, and produces decisions based on that information.[8] The neurons are the basic unit of information processing. Each neuron takes in an input, processes the input, and then produces an output.[8] Working together, a network of neurons can produce a more complex output.

Typically, the neurons are organised into layers based on their function.[8] The topmost layer is the input layer, with neurons that receive information before passing it on to the second layer, known as the hidden layer.[8] In this layer, the information is processed and passed on to the final layer, the output layer, which produces the requested final output.[8] The process of passing from layer to layer is stimulated by an activation function. The activation function is like a switch. When the switch receives an input that tells it the neuron has completed its function, it passes the output to the receiving neuron in the next layer, which receives that output as an input.[8]

A neural network can have many hidden layers that process information multiple times in multiple ways; when this happens, it creates the deep learning machines we have discussed. The connections from neuron to neuron, from layer to layer, are also assigned weights.[8] As a neural network learns, it can automatically adjust the weights of different connections to favour using connections that produce the correct answer.[8]

As with the human brain, an artificial neural network can be trained. Let's return to our dog photo example. Think about how you learned what a dog is. You probably saw many images of dogs or dogs in real life, and somebody told you to start associating their basic physical characteristics with the word "dog." Neural networks undergo a very similar process. To train a neural network, a trainer first gives the network several input examples and the correct output based on those examples.[8] The neural network learns by comparing the answer it gives to the correct answer and adjusting the way it weights certain neurons.[8] In our dog example, after receiving the image inputs and correct identifications, the network may still produce the response "cat" or "George Clooney" during its first test. However, the network will then see that this is incorrect and will run the input-analysis-output process again and again until it produces the word "dog." As the training process continues, the network will be able to produce the correct answer with increasing speed and accuracy.

Neural networks have become the cornerstone for much of the AI that we use in our daily lives. Here are just some of the ways that neural networks are already impacting you:

- **Speech recognition.** Voice assistants like Siri and Alexa use neural networks as they train to recognise and respond to human speech.[9] They convert spoken language into text and then analyse the text to determine an appropriate response.[9]

- **Recommendation systems.** Any time you pull up Netflix and use the "Recommended For You" section to decide what to watch, you're benefiting from neural network learning.[9] Platforms like Netflix and Amazon use your actions and the actions of similar users as inputs for a neural network to produce personalised recommendations.[9]

- **Self-driving cars.** The advent of self-driving technology is a result of engineers' ability to train self-driving software to recognise real-time situations.[9] The software utilises cameras and radars to take in image information, interpret what is happening on the road, and react to those stimuli.[9]

- **Smartphones utilise various types of neural networks.** Facial recognition technology, autocorrect, and dynamic battery life are just a few examples of how your smartphone learns and adapts to your behaviour.

- **Medical technology.** An extension of image recognition technology, AI can take in inputs from CT scans, X-rays, or MRIs and detect patterns or abnormalities that a human might miss.

- **Financial forecasting.** Neural networks are used in stock market predictions, fraud detection, and risk assessment. They analyse huge amounts of data to understand trends, which gives useful insights to investors and helps financial institutions identify fraud by flagging a transaction as being outside of a trend.

The Different Types of AI Explained

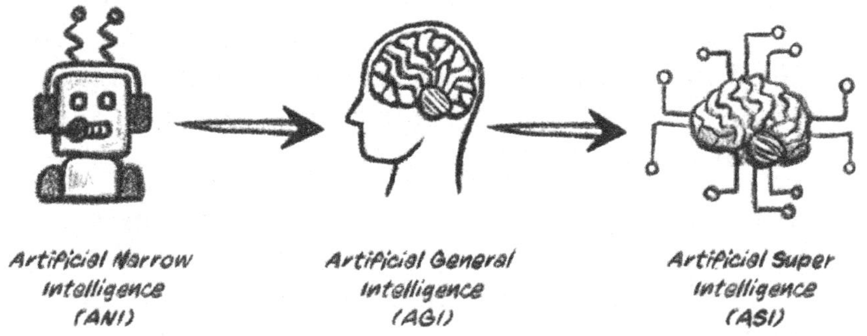

Not all artificial intelligence is built the same. Some AI is built to fulfil a specific objective, while other AI is built for more general purposes with the goal of mimicking human intelligence. Researchers generally define three different categories of AI: narrow AI (AI designed to do a specific task), general AI (AI designed with a human-like intelligence quality and able to undertake a variety of tasks), and super AI (AI with such a high degree of intelligence that it surpasses human intelligence).

Narrow AI

At present, this is the only kind of AI that exists in anything other than theory. All the AI that you know, including voice recognition systems like Siri and Alexa, chatbots like ChatGPT, and self-driving car software, is considered narrow AI.[9] Narrow AI can be further broken down into two subcategories: Reactive machine AI and limited memory AI.

A reactive machine AI performs narrower tasks than a limited memory AI because it does not store information in its memory.[7] It only uses the presently available data to make decisions and produces intelligent results by analysing a vast quantity of data at one time.[7] Examples of reactive machine AI include:

- **Streaming service recommendations.** Your recommendation neural network doesn't store your catalogue of watched TV shows in its memory. When you pull up Netflix, the neural network accesses Netflix's stored log of your watched content, then compares it to the logs from other similar consumers, and puts out a recommendation list.[7]

- **IBM's Deep Blue.** The chess-playing supercomputer famously beat grandmaster Garry Kasparov in 1997, but it did so without maintaining a memory of past gameplay.[7] Deep Blue does not use past games or past mistakes to inform future decisions; it simply analyses the current game situation and makes the best move based on a constellation of predictions about what could happen next.[7]

As the name suggests, limited memory AI maintains a store of past events and interactions to inform future outputs.[7] This type of AI is closer to what our brains are capable of than reactive machine AI.[7] We are not capable of analysing massive amounts of data or visualising a huge web of potential outcomes, but we can remember information for a limited amount of time and use that information to make informed decisions. However, unlike the human brain, limited memory AI cannot store information in a long-term memory library.[7] In limited memory AI, the AI draws from a store of short-term memory in conjunction with present information to produce an answer, replacing the previous information stored in short-term memory.[7]

Some examples of Limited Memory AI include:

- **Chatbots and Generative AI.** AI that processes and responds to human voice or text, like ChatGPT, Siri, and Alexa, stores limited information about previous interactions to determine an appropriate response to the user input.[7]

- **Self-driving cars.** Self-driving software stores information about the world around it in limited memory, combining real-time information about other cars' speed and direction with knowledge of upcoming traffic lights or curves in the road to produce a driving output, like having the car slow down or change lanes.[7]

General AI

Currently, general AI exists only in theory, but it is worth offering a brief description of what it is and how it will work. General AI, also known as strong AI, is designed to mimic human intelligence by storing lots of information and by being able to understand and interpret lots of different tasks.[9] General AI will be able to analyse unfamiliar problems by drawing on past experiences and assigning meaning to those experiences.[9]

Super AI

We may someday reach a point where AI has become so advanced that it surpasses human capabilities. At such a time, this AI will be known as super AI. Because this AI will be able to do things humans cannot, it is impossible to understand the potential ramifications of its existence. In theory, super AI will not only be able to understand human sentiments and emotions, but it will also be able to experience emotions, needs, ideas, and beliefs of its own. Some researchers suggest that we will never be able to create such an AI, as creating a neural network that surpasses human capabilities seems like a logical impossibility. Others believe that we should not try to create such technology, as we cannot understand the risks that its existence will pose. However, at present, the march of AI seems inevitable, and only time will tell if superhuman intelligence is possible.

Summary of Chapter Learnings

Artificial intelligence is designed to take in an input, process that input, and return an output.[7] This seems like a simple goal, and it is, but it's that simplicity that makes AI so powerful. An AI machine can be taught to do a near-infinite amount of things, limited only by the interests and needs of its programmers.

AI evolved from years of research on computers and programming, but it was the advent of deep learning in the early 2010s that jump-started the current AI revolution. Deep learning is the product of a neural array known as a neural network.[8] A neural network is similar to a human brain. The artificial neurons take in information and communicate with other neurons to produce a desired output.[8] Within a neural network, the neurons are organised into three layers: the input layer, with neurons that take in a user input; one or more hidden layers, which take information from the input layer and process it; and the output layer, which takes information from the hidden layers and uses that information to produce the desired output.[8] When a neural network has lots of hidden layers, it has added computational power and, therefore, can perform increasingly complex tasks.[8] When an AI machine has multiple layers, it is said to be engaged in deep learning.[8]

The phrase "deep learning" is significant because it highlights the importance of AI's ability to learn. The strength of artificial intelligence lies in its capacity to be trained to generate accurate results for a given query. This is done by training a neural network with a set of inputs and their corresponding correct outputs.[8] When presented with a test input, the network attempts to produce a correct output. If its output is incorrect, the network tries different neural pathways until it achieves the correct result. The successful pathway is then assigned a higher weight to ensure it is used in future queries.[8]

Three different categories of AI are theoretically possible, though only one is currently achievable. This is the weakest category of AI, known as narrow AI. Narrow AI is so called because it has a narrow function; it can only complete a single, specific task, like answering a question or providing TV show recommendations.[9] Narrow AI is divided into two categories based on whether it stores information in a memory bank. A reactive machine AI only uses the present information and does not store data in a memory bank.[9] It gains its intelligence by comparing massive amounts of data that exist in a given moment to produce an output.[9] Limited memory AI utilises a form of short-term memory. It incorporates both past and present data to generate an output. Two other AI types are still being theorised. These future AI types would either have the ability to perform various tasks like humans or potentially even surpass human abilities entirely.

Understanding how AI works will help you determine what style of AI best suits your workplace needs. In the following chapters, we will dive into more detail about how you can use your newfound knowledge of AI's function to bring it into your workflow.

3
ENHANCING PRODUCTIVITY WITH AI

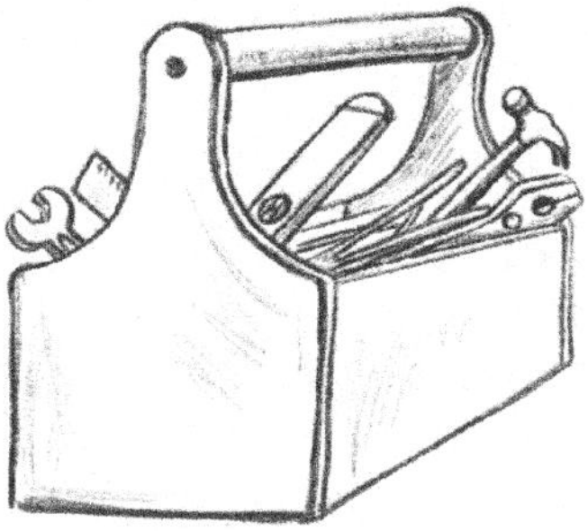

Now that you have a working knowledge of AI's function, it's time to take action! In this chapter, we will look at a variety of options you have at your disposal to bring AI into your workflow. We're going to dive deep into seven areas of your workflow that you could start automating right now, and we will examine some specific tools and applications that you can explore. Start visualising a smoother, easier workday, and let's get into it!

1. Content Creation and Curation

One of the easiest ways to start incorporating AI into your workflow is to have it draft certain types of content for you. Consider a scenario where you need to give stakeholders an update on the progress of a project plan. Writing such an email manually can be a meticulous task that requires a considerable amount of time. This time could be better spent on analysing the project and refining strategies. This is where a conversational AI tool, like ChatGPT, becomes immensely helpful. By instructing the AI to create an initial draft detailing the project's current status and next steps, you can get a strong foundation for your communication. This process allows you to quickly personalise and edit the message before sending it out, thus optimising your time for critical project management activities.

There are various applications of AI in content creation. For instance, you can input data into a chatbot and request it to analyse the data based on specific parameters and generate a report. Additionally, you can use an image generation tool to design a logo for your company or create a poster for marketing campaigns.

AI Toolkit - ChatGPT

Each productivity strategy comes with an AI software toolkit to help integrate AI into your workflow. Here are some resources you can use to automate content creation and curation:

- **Writing assistance:** OpenAI's ChatGPT is probably the most well-known tool today, and for good reason. Its ability to take on almost any text-based task quickly and accurately is incomparable. ChatGPT alternatives include ClickUp, Writesonic, and Copy.ai.[13]
- **Graphic design:** Tools like DALL-E or Canva's AI generator can create and suggest design elements.[13]

Example

Each productivity strategy will also show an example of its deployment in a real-world scenario. Here, the scenario is a study conducted on the effect ChatGPT has on productivity for mid-level professional writing tasks.[13] In the study, participants were assigned occupation-specific writing tasks, and half were asked to use ChatGPT to complete the task.[13] The study found that using ChatGPT lowered the time taken to complete the task and raised overall task quality.[13] The researchers also extolled the value of ChatGPT to reduce worker inequality and increase job satisfaction, as employees spend more time on idea creation and less time on rote writing tasks.[13]

2. Data Analysis and Interpretation

If you have a large amount of data, you can utilise AI to build predictive models that can efficiently process vast amounts of data and provide you with valuable insights. AI has the ability to analyse data without bias, and it can handle large amounts of data much faster than humans.[14] Additionally, AI can sort your data automatically, saving you the trouble of organising your data.[14]

AI Toolkit – Polymer

- **Automated data processing:** Polymer is a software that can perform data analytics by processing and interpreting large datasets and undertaking predictive analysis.[14] Other software includes Tableau, MonkeyLearn, and Microsoft Power BI.[14]

Example

An e-commerce business streamlined its marketing data analysis by using Polymer, an AI tool that integrates data from platforms like Google Ads, Shopify, and Facebook.[14] Polymer's AI-driven dashboards and natural language processing enabled the team to save hours on manual data collection, quickly adjust underperforming campaigns, and improve ad conversions by 19% in days.[14] This automation enhanced decision-making and collaboration across departments, driving better overall marketing performance.

3. Personalised Customer Engagement

You can use AI to streamline your marketing strategy. For example, you could build a data analysis tool to analyse consumer data and discern trends, or you could build an AI network that creates personalised marketing campaigns (not unlike how Netflix and other streaming services use custom recommendations).[12] You could also have AI write social media posts based on consumer data or put a chatbot on your company's website to answer user questions.[12] Again, the name of the game here is personalisation. The beauty of AI is that it can tailor a marketing campaign to an individual rather than more traditional demographic-based marketing strategies.

AI Toolkit - HubSpot

- For AI in marketing, sales, and content creation, you can use HubSpot, a service that manages personalised marketing content and strategies.[12] For maximum efficiency, consider combining a marketing management tool like HubSpot with a data analysis tool.

Example

Bloggers and SEO writers have found HubSpot to be particularly helpful for their work. According to a survey conducted in 2023, 84% of bloggers believe that AI has been beneficial for their marketing strategy.[12] Bloggers have used HubSpot to generate content creation strategies, optimise their websites for SEO, and enhance the user experience.[12]

4. Automation of Routine Tasks

Are there certain tasks that take up a disproportionate amount of your workday? You can probably use AI to automate those tasks. For example, you can have AI handle meeting scheduling, invoicing, and file organisation. You can also use AI for process optimisation, having AI analyse your workflow to identify ways to improve efficiency or spot areas for potential human error.

AI Toolkit - Asana

- **Workflow automation:** UiPath offers Robotic Process Automation (RPA) tools to automate repetitive administrative tasks.

- **Process optimisation:** Company workflow programs like Trello or Asana integrate AI to optimise project management and workflow processes. If your company already uses workflow software like Trello or Asana, experiment with AI integration to see what it can do for you.

Example

Asana is a project management tool that incorporates AI to help you manage your projects more efficiently. With Asana's AI, you can automatically analyse tasks and project timelines to determine the most critical tasks. Additionally, it can suggest which team members should be assigned to a given project based on their experience and past performance. Asana's AI can also predict where bottlenecks may occur in the workflow, allowing you to adjust deadlines based on which parts of the project are likely to take the most time. With Asana's AI-powered features, you can streamline your project management process and maximise your team's productivity.

5. Advanced Research, Development, and Creativity

AI is not limited to performing routine tasks that one may not want to undertake. It can also be helpful in the process of idea generation and innovation. Particularly in areas like engineering and technology, AI can help build models to test theories without creating a physical prototype. It can also contribute to the innovation process in software development by generating new code.

AI Toolkit – DeepMind

- **Accelerating innovation:** AI platforms like DeepMind are useful for research and generating innovative solutions.
- **Simulation and modelling:** ANSYS is a simulation software that employs AI to enhance its modelling capabilities.

- **Coding:** OpenAI Codex is a tool that interprets natural language commands and turns them into code. It can write in more than a dozen programming languages.
- **Idea generation:** Miro uses AI to facilitate brainstorming sessions and suggest new ideas.
- **Design and prototyping:** Autodesk's generative design software leverages AI for rapid prototyping in architecture and product design.

Example

DeepMind has been used to build football models, analysing players' behaviours in different positions during different scenarios like penalty kicks.[15] Researchers believe these models could help democratise the football industry by automatically selecting interesting clips from the game to serve as highlights.[15] This will eliminate human bias toward choosing highlights from the game's most famous players, allowing lesser-known players to be recognised.

6. Improved Decision Making

AI can create simulations and analyse business scenarios to aid decision-making. It can also conduct risk assessments to evaluate different scenarios for their relative level of risk and strategise accordingly.

AI Toolkit - Tableau

- Scenario analysis: Tableau's analytics technology can be adapted for businesses, creating visualisations of different business scenarios.[14]

Example

AI has found a foothold in the insurance industry, as it can assess an insured's level of risk. For home insurance, AI can combine factors like the home's location, age, and value to determine the risk of damage and the quantity of money the insurance agency is likely to lose if damage occurs.[11] The AI can also suggest mitigation measures, like installing a sump pump or reinforcing the roof.[11]

7. Collaboration and Communication

If you work for an international company, AI can facilitate language translation between employees from the countries where your company operates. It can also streamline communication by automatically taking meeting notes and scheduling follow-ups. Suppose you have multiple people working on one project or one document. In that case, AI can enable multiple users to work on the project simultaneously in real-time, track changes to that document, and provide feedback.

AI Toolkit – Otter.ai

- **Smart meeting assistance:** Otter.ai transcribes and summarises meetings, aiding communication by ensuring that everyone has a record of a meeting's takeaway messages.
- **Language translation:** Software like Microsoft Translator or the SayHi app uses AI for real-time translation.

Example

In the legal field, AI-powered language translation tools translate case documents and supporting notes from languages that legal experts may not be familiar with into their native language. This makes it easier for them to access important information that could be helpful in their case.

Exploring the Results of Advanced Productivity

As we've seen from the above examples, AI has near-limitless possibilities in the workplace, from automating menial tasks to assisting in idea generation and innovation. All of these uses connect back to a central benefit: increased productivity. No matter how you choose to deploy AI in your workplace, it will indeed allow you to produce better work faster.

Here are just a few of the benefits you will draw from increased productivity in the workplace:

- **Work-life balance improvement:** Improving work-life balance is possible by automating time-consuming tasks, which allows you to focus on the more engaging parts of your job. As a result, you will spend less time on your job overall, leading to increased job satisfaction and less stress outside of work. Additionally, reducing the time spent on work will free up more time to devote to your personal life.
- **Enhanced creativity and problem-solving:** Many employees find that working within a team stimulates creativity. Using AI to help develop new ideas is like having a virtual brainstorming team. Employing AI to assist with the idea generation and innovation process can jump-start your idea flow, leaving you feeling creatively stimulated and equipped to tackle complex problems.
- **Reduced work stress:** The efficiency brought by AI in managing routine tasks reduces the overall stress associated with tight deadlines and high workloads. With AI, you can focus on high-value tasks without being overwhelmed by the volume of work.
- **Improved decision-making:** With a litany of valuable data and analytics at your disposal, you can make informed decisions free of human bias. This will empower you to confidently make decisions, thereby boosting feelings of security and satisfaction in your role.
- **Enhanced communication and collaboration:** Tools like automated meeting notes and real-time translation enable you to collaborate efficiently with coworkers. In

particular, these tools allow you to build an international team, bringing in more diverse perspectives, broadening your professional network, and enhancing teamwork skills.

- **Adaptability and resilience:** A working knowledge of AI equips you with the flexibility to respond quickly to an emergent situation. It also prepares you to be valuable in a workplace where technology is changing job demands almost daily.
- **Risk awareness and mitigation:** If you've ever laid awake at night worrying whether your big project will fail, AI risk assessment tools can help you sleep easier. Risk assessment tools can help you identify potential issues before they happen, and they can suggest mitigating measures, reducing the likelihood of project failures or setbacks.
- **Improved time management and efficiency:** If you're a natural procrastinator or worried about working under tight deadlines, you can use AI to set manageable goals for yourself and optimise your workflow. This will give you more control over the structure of your workday.
- **Increased recognition and opportunities:** As mentioned in Chapter 1, most companies now recognise the potential for AI to transform the workplace, and three out of four companies are specifically looking to hire AI-savvy employees. If you can become an AI-savvy employee, you will garner increased recognition within your company. Alternatively, if you want to climb up the career ladder or make a career change, AI skills will help you be competitive in your search.
- **Financial growth and earning potential:** With greater career recognition comes increased earning potential through productivity incentives or career advancement. You can even use AI to optimise a side hustle, earning you some extra money without having to devote too much time to your additional work.

Summary of Chapter Learnings

AI has near-limitless potential to revolutionise your workflow; the only limit is your imagination. At its most fundamental level, AI can automate repetitive tasks that consume a disproportionate amount of your day, such as email correspondence, meeting scheduling, and data entry. However, this is just the tip of the iceberg when it comes to AI. AI can assist you in creating or curating content, generating graphic design suggestions, writing reports, and more. You can use it to learn new skills, conduct research, and perform complex analytics. Additionally, AI is beneficial for large companies with a global workforce, as it facilitates collaboration with language translation and automation tools.

The overarching benefit of AI is increased productivity. All these AI strategies are designed to help you do better work and be more efficient. The AI revolution goes hand in hand with the "work smarter, not harder" mentality. Bringing AI into your workflow can reduce workplace stress by freeing up your time to focus on the parts of your job that you enjoy or spending less time on your job overall, promoting work-life balance and increasing job satisfaction. Even within the best parts of your career, AI can assist you as a brainstorming partner, generating suggestions to stimulate your creativity and help you solve challenging problems. It can provide you with a wealth of information to help you make informed decisions and manage risk. Finally, it can help you be flexible in a dynamic situation and prepare you to succeed in a workplace where technological advancements are seemingly inevitable.

4
OVERCOMING AI CHALLENGES AND FEARS

Hopefully, you now see that AI has tremendous potential to revolutionise the way you work. That said, you might still be worried about what it will mean to incorporate AI into your workflow. Perhaps you've read reports disparaging AI as a harbinger of job loss or that we're in the nascent stages of a Matrix-esque world where a computer runs everything we know. Perhaps you're interested in AI's potential but think the learning curve will be too steep for you to manage. Well, this chapter is for you! In this chapter, we will dispel some of the myths and misconceptions that surround AI.

The AI Employment Paradox: Navigating Job Security in the Age of Automation

Artificial intelligence is taking on increasing responsibility in the workplace, including responsibilities that were once the domain of human employees. Certain analysts believe that we are currently in the early stages of what is known as the "AI employment paradox." This phenomenon occurs when businesses use AI to automate certain job functions, eliminating certain positions previously filled by humans and potentially resulting in job losses. These analysts caution that, if left unchecked, the AI employment paradox will result in a work environment where only highly skilled employees are useful and low to middle-skill employees cannot find work because their jobs have all become automated.[19] This is a widespread concern among employees: a recent study revealed that one-third of workers are scared that AI will replace their jobs in the next three years.[22]

There is likely some truth to the employment paradox; a certain amount of job loss will be inevitable as low to middle-skill jobs become automated. However, rather than using this truth as a soapbox to criticise the AI revolution, we should take this inevitability as an opportunity to reconsider what skills we value as employees. In an AI-dominated workplace, soft skills like critical thinking, creativity, and empathy will become the hallmarks of a successful employee, as these skills are still distinctly human. Additionally, interdisciplinary skills will become essential. Having the ability to work at the intersection of fields like business, technology, and ethics will allow you to take a holistic approach to implementing the results of AI automation. As an employee, it is worth taking some time to consider your skill set. Do you have soft skills that you can use to complement AI's automation capabilities? Can you not only operate an AI machine but also interpret the results that it produces?

Current Impact of AI on Jobs

A common concern among critics of AI is that it will lead to significant job losses in certain sectors, potentially causing them to shrink or disappear entirely. Humanities and arts jobs are often said to be on the chopping block, as are data analysis and retail jobs. Even jobs that experts have long held to be "safe" positions like coding or language translation are said to be under threat as AI becomes increasingly capable in those areas.

The reality is that it's unlikely that any one sector will face a dramatically greater impact than another. The more likely possibility is that each sector will have certain types of jobs reassigned to AI. Unsurprisingly, most experts believe that AI will replace jobs that are easily automated or that require little critical thinking.

However, AI may only partially replace human workers, even for these low to middle-skill jobs. For example, customer service workers are prime candidates for job replacement, as companies increasingly use chatbots to answer customer queries and address complaints. However, the present implementation of AI in customer service roles shows signs of enhancing worker productivity rather than entirely replacing the worker.[18] In their current implementation, chatbots can handle routine requests, freeing up human workers to deal with more complicated cases. Before passing a case to the representative, AI-powered assistants can collect relevant information from the customer, allowing for more efficient case management.

The customer service representative example demonstrates the likelihood that AI will lead to more job transformation than job reduction. This means that instead of being entirely replaced by AI, job roles will evolve to complement AI capabilities. We've already examined some of the ways that this change might unfold. Insurance agents may spend less time assessing risk and more time interpreting results and making client decisions. Bloggers may spend less time researching SEO optimisation and more time creating valuable content for their readers. This not only saves them time but also allows them to create content that is more engaging and relevant to their audience. All these job transformations demonstrate how crucial critical thinking skills and emotional intelligence will become.

Job Creation Through AI

Some jobs will inevitably be lost due to the rise of AI technology; however, it is essential to note that new job opportunities will also be created. While AI can perform many tasks, humans still possess superior skills in generating new ideas and interpreting outcomes. Therefore, new jobs will likely require employees who possess critical thinking skills, creativity, and innovation.[19] Many of these jobs will involve working alongside AI technology to maximise efficiency. As AI becomes ubiquitous in the workplace, we will need people who understand what it can currently and possibly do.

Integrating AI technology will require diverse human workers to perform various AI-related jobs. The most obvious among them are the engineers who develop algorithms and models that form the basis of AI. Once the AI system is implemented, employers will need workers to maintain it and ensure it functions as intended. Workplaces will require individuals to train others on AI usage, research new ways to implement AI and ensure ethical usage.[16] Additionally, companies that develop AI products for the workplace will need sales professionals who understand their products, the marketplace, and individual client needs. These are the only jobs that we can conceive of based on what we know about AI right now. As AI evolves, new job sectors that we haven't yet imagined will also arise.

Skills For the Future

The AI revolution may not be a harbinger of mass layoffs and staggering unemployment problems. Still, it will create a workplace environment where you will get left behind if you do not learn how to incorporate it into your workflow. It's important to start forming a habit of practising the skills that are essential for a world centred around AI. In particular, three skills stand out as being the hallmarks of an AI-savvy employee: digital literacy, critical thinking, and adaptability.[16]

- Digital literacy: It's hard to use AI if you don't know how to use AI. Reading this book is a great first step towards learning how AI works and how to incorporate it into your workflow. Once you've finished the book, return to the previous chapter, read through the AI Toolkit lists again, and start dabbling in some of the discussed software. The easiest way to learn to use something is to start using it.

- Critical thinking: AI is capable of producing a wide range of valuable information, but you will still need to possess the skill of combining that information. You can practise this skill within your own company by analysing a decision that your company has recently made. Look at some sample data and try to arrive at the same conclusion your company did.

- Adaptability: Given how fast AI technology is advancing, you must be prepared to learn new skills and technologies constantly. You must also learn how to adapt your workflow to incorporate your newfound expertise.

Debunking AI Myths: Separating Fact from Fiction

The AI employment paradox might be the most prominent myth currently swirling around AI, but it's certainly not the only one. In this section, we will debunk other myths many people believe about AI.

Myth: AI Will Fully Replace Human Intelligence

This one is for the sci-fans, the folks who are worried that we're not far off from a real-life HAL 9000. As discussed in Chapter 2, such an AI would be known as a General AI or a Superintelligent AI, and as we know, these are both still very much theories. There is immense speculation as to whether we can even build such an AI, and even if we could, there will likely be huge discussions about whether we should. Needless to say, HAL 9000 and its human-like counterparts are nowhere near imminent.

Myth: AI is Always Completely Unbiased and Infallible

In most cases, AI performs its tasks with high objectivity and accuracy, often more so than a human could. However, AI is still evolving, and most AI systems are still being trained and refined.[17] Errors can come from various places, including incomplete data, algorithmic limitations, or edge cases that fall into a gap for which AI is not prepared.[17] AI can also be subject to occasional bias, usually resulting from bias in the training or the algorithm.[17] This is why human workers are still needed to monitor, train, and refine AI.

Myth: AI is Only Useful for Large Corporations

While large corporations may have more resources to invest in deploying more sophisticated AI, it is certainly not their exclusive domain. Workflow tools, automation resources, and content creation software are already accessible to small businesses, and AI technologies across the board are becoming less and less expensive.[21] In particular, AI is now being provided as a cloud service, which allows businesses to scale down the AI system to meet their needs.[18] It also provides remote workers with access to AI tools and reduces the need for on-site AI IT services.[18]

Myth: AI Is Such a Massive Revolution That We Have to Take it Slow

One of the best things about AI is that it can scale very quickly, making it straightforward to deploy for widespread use across a business. If AI has the potential to radically transform your business from Day One, why implement it in tiny drops? Of course, this doesn't mean you should suddenly automate everything you do. It's worthwhile to start by automating just a few parts of your workflow, but if you find it proving to be a successful experiment, be prepared to scale it quickly.

Myth: AI is Risky Because it is So New

As we've seen with examples like facial recognition and speech-to-text technology, AI has found a home in many different parts of our lives, and if you stop and think about these technologies, they've been around for quite a while. What makes the AI revolution seem new is that it has only recently matched power with the kind of accessibility that appeals to businesses.21 The developers that built the technology you are starting to see in your workplace did not start from scratch; they built upon years of experience and refinement to reach the point that we're at now.

Understanding AI's Limitations

We have touched on this fact, but it bears repeating in its own section. AI currently possesses the power to transform the workplace, but it still needs improvement. Bias and fallibility can still creep into the results that AI produces. Furthermore, AI is still unable to fully replicate the qualities that define human thinking, such as critical thinking, decision-making power, and creativity. However, consider reframing these limitations as an opportunity. As an employee, you have the power to integrate yourself into the workplace spaces that AI can't fill. Find ways to tap into your innate ability to build, engineer, and pioneer to become an essential employee that complements your workplace's AI for both the short and long-term future.

Summary of Chapter Learnings

Understandably, a lot of people are afraid of AI. So many misconceptions exist about its origins and implementation in our world, and it's easy to be scared of things we don't understand. For example, the most common misconception surrounding AI in the workplace is that it stands to cause massive job loss as it takes on responsibilities formerly assigned to human employees. This is known as the AI Employment Paradox, and it has received much media attention in recent years. There is likely a bit of truth to the AI Employment Paradox, especially because employees who either struggle to integrate with AI or outright refuse to do so will be left behind in AI's march towards ubiquity.[19] However, AI is doubtful to cause mass job layoffs and leave vast swaths of the workforce unemployed. There are two reasons for this. First, the AI revolution will likely lead to more job transformation than job reduction. This just means that, rather than AI replacing an employee entirely, the employee's job description will change to complement the AI's work. Second, even if some jobs are lost to AI, new jobs will be created. Many of these jobs will relate to the AI itself. We will need people to build new AI modules, train the AI, teach other people how to use AI, sell AI technology to prospective users, and create regulations based on the ethical use of AI. Both job transformation and job creation will reward those employees who figure out how to integrate themselves into the spaces that AI cannot yet fill. This will require digital literacy to know how to use the technology, critical thinking to interpret the results of AI integration, and adaptability to learn and integrate new technology as it becomes available.

The AI Employment Paradox is not the only myth associated with AI. Some people believe that AI is on the verge of gaining human intelligence or that AI is too expensive, new, or risky to be widely useful. However, AI with human intelligence is still very much theoretical, but the technology we do have is getting increasingly accessible, mainly through integration with cloud services.

It is also important to dispel the myth that AI is a perfect end-all solution for acquiring infallible, objective information. AI is built and trained by humans, and humans are naturally fallible and biased; therefore, AI will reflect a certain amount of this reality. However, despite the concerns about the limitations and biases of AI, it can still perform its functions with remarkable accuracy. With some human guidance and intervention, AI can help us identify and eliminate our own biases. This, in turn, can make our decision-making processes more efficient and informed.

5

TRANSFORMATIVE TALES: CASE STUDIES IN AI-DRIVEN EMPLOYEE EMPOWERMENT

The transformative power of AI might sound great in theory, but you might be asking: does the theory hold up in practice? How are employees actually being impacted by the concepts we've discussed so far? In this section, we'll examine some case studies from companies that integrate AI into their workflow and how that integration impacts their employees' work.

Amazon Go Stores Revolutionise the Retail Experience

If you've never been to an Amazon Go store before, you're missing out on a shopping experience unlike any you've had before. All Amazon Go stores are cashierless, using AI technology like computer vision, sensors, and deep learning algorithms to automate the checkout process. As a customer, all you have to do is walk in, pick up the items you want, and leave, and the AI will automatically charge a credit card that you have on file with the company. At first glance, this might seem to contribute to the AI Employment Paradox by eliminating one of the jobs long considered essential to retail establishments. Amazon Go stores still employ many employees, but these employees are no longer needed to work at checkout counters. This frees up their time to manage stock, maintain the AI infrastructure, and respond more quickly to specific customer requests.

ChatGPT Improves Efficiency in the Consulting Industry

A 2023 study examined the impact of ChatGPT on daily workflow at the Boston Consulting Group, a large management consulting firm.[24] In the study, 758 consultants from the firm were first assessed for a baseline level of productivity.[24] They were split into three groups: no AI access, ChatGPT-4 AI access, and ChatGPT-4 access with some training on how to use it.[24] For those tasks that were completed using AI, ChatGPT users were significantly quicker, completing 12.2% more tasks and completing each task 25% faster.[24] They were judged to have produced 40% higher quality material.[24] Consultants whose baseline productivity was below average saw productivity rate increases of 43%, while consultants with an above-average baseline saw productivity rate increases of 17%.[24] These results suggest that ChatGPT has the potential to bring lower-performing workers closer to the median, thereby levelling the playing field between lower and higher-performing employees. It also seems that ChatGPT can substantially increase the quality of work being performed for those tasks with which the program can assist.

Siemens AG Embraces AI for Project Management and Forecasting

Siemens AG is a global technology company specialising in large-scale projects like infrastructure development. Recently, they have begun using AI algorithms to predict project delays, forecast budget overruns, and identify potential resource shortages.[27] Siemens has declared their AI integration to be transformative. Project delays have been reduced, budget overruns have been minimised, and resource allocation has become more precise.[27] These transformations have allowed Siemens to optimise the use of its skilled workforce, enabling them to focus on their jobs without wondering whether their project will go over budget or whether they will have the resources to complete their project.

Uber Helps Drivers Maximise Earnings with AI Algorithms

Drivers for Uber, the international ride-sharing service, may have noticed AI-influenced improvements in pay and driving routes. For example, based on historical and real-time data, Uber's AI technology can suggest the most profitable times and locations for passenger pick-ups to drivers.[28] Additionally, real-time traffic information helps drivers select the fastest route.[28] These changes have allowed drivers to incorporate fact-based decision-making into their choice of driving location and route rather than relying solely on experience and instinct.

The Construction Industry Experiences Less Down Time Thanks to Predictive Maintenance AI

One of the biggest hurdles to on-time construction project completion is equipment breakdowns; when construction workers don't have access to heavy machinery like bulldozers and dump trucks, they aren't physically able to complete their work. But now, companies like Caterpillar are installing predictive maintenance AI on their machines. This AI can take in information on equipment temperature, vibrations, and usage patterns to recommend a maintenance schedule. Installing predictive maintenance, AI has significantly reduced Caterpillar's equipment breakdowns, decreasing unexpected days off due to the company's equipment.[26]

McDonald's Uses AI to Personalise the Customer Experience

If you've ever found yourself craving McDonald's chicken nuggets on a rainy day, you might have new AI personalisation technology to thank. McDonald's has adopted an AI personalisation platform to adjust menu displays and recommended items based on historical ordering and real-time data, as well as the time of day, season, and even daily weather.[29] This means that McDonald's employees who are responsible for managing their restaurant's stock must adapt to new information about customer preferences and adjust their stock orders accordingly.[29]

Fostering the 'AI Integration Flywheel'

If you've never heard of a flywheel, here's a quick lesson. A flywheel is a device that is used to store rotational energy. They have been used for centuries to create machines like the potter's wheel, which rotates constantly to help potters shape their pots. When at rest, flywheels are hard to start turning because they experience inertia, meaning they have a physical tendency towards staying at rest when they are at rest. However, once the flywheel gets going, it will tend to stay in motion. Furthermore, if consistent force is continually applied to the flywheel, its speed and energy will increase, thereby increasing its momentum and making it harder and harder to stop.

The flywheel analogy is easily applied to a business context. For a business, the flywheel represents the interconnected set of activities and processes that drive success. Typically, these activities connect in a circular pattern. For example, producing high-quality work results in higher customer satisfaction and retention, leading to increased profits that can be invested back into producing higher-quality work, creating a cycle of success.

"The flywheel model is a framework that explains how large companies are created through continuous and accelerating efforts." Take Amazon as a quintessential example of a flywheel success story. From the outset, Jeff Bezos was intent on customer satisfaction and low prices being the driving force to get his company going. Initially, selling products at low prices meant smaller profits. However, low prices combined with a focus on customer service got his business flywheel spinning. Once Amazon perfected the art of customer retention, its flywheel started spinning faster and faster, eventually turning it into one of the world's largest companies.[23] Today, Amazon still focuses on customer service as the continual influx of force that keeps their flywheel spinning.[23] Initiatives like Amazon Prime, flexible return policies, fast delivery, and a convenient shopping model help drive traffic to Amazon sellers, who then provide a satisfying customer experience that leads to more consumer traffic.[23]

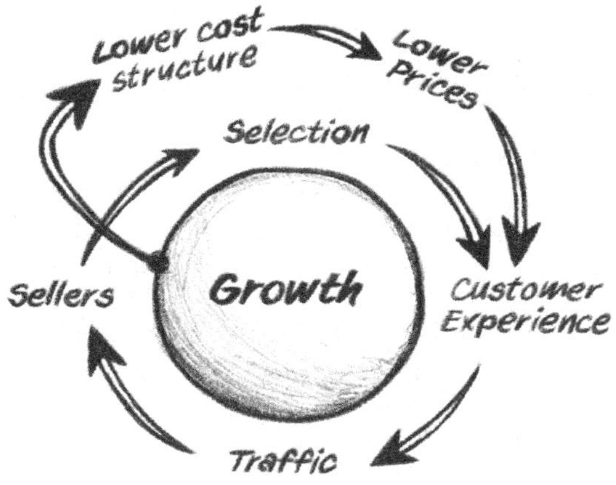

(Amazon Flywheel)

As an employee, you can adopt the flywheel model for your own personal development as you strive to become an AI-savvy employee. To start the wheel moving, introduce AI integration into your workflow, whether by automating some of your work or using it to help with content creation or decision-making. AI integration will spin the wheel towards increased productivity. Then, as your productivity increases, the wheel will spin towards increased opportunities, both in your professional life (as increased productivity will make you a more valuable employee) and your personal life (as you will free up more of your time for personal pursuits). The simultaneous increase in personal and professional opportunities will contribute to a heightened sense of confidence and fulfilment in your current role. Finally, your newfound confidence will demonstrate how AI integration meaningfully impacted your personal and professional growth, and it will lead you to introduce more AI into your work, thereby feeding the flywheel's momentum.

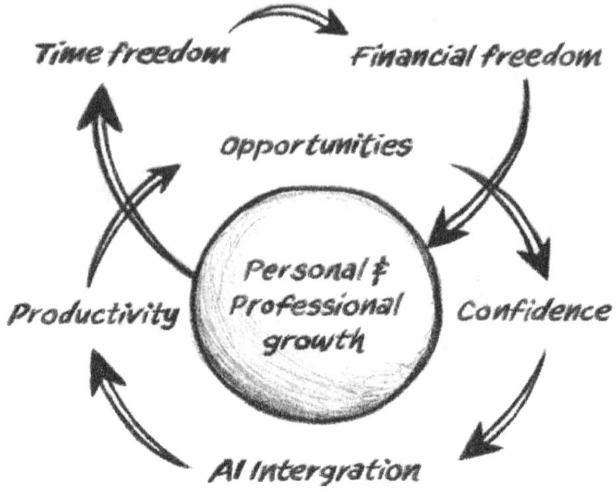

(Your AI Integration Flywheel)

At first, the productivity, opportunity, and confidence gains might be small because you will be starting a new process; you will be moving the flywheel from its resting state, and remember, it takes a lot of effort to start a flywheel! But once you notice the benefits, even if they're not huge, keep feeding AI integration into your flywheel, and it will start spinning faster and faster. Eventually, you will have an unstoppable personal and professional growth flywheel that will yield significant productivity, opportunity, and confidence.

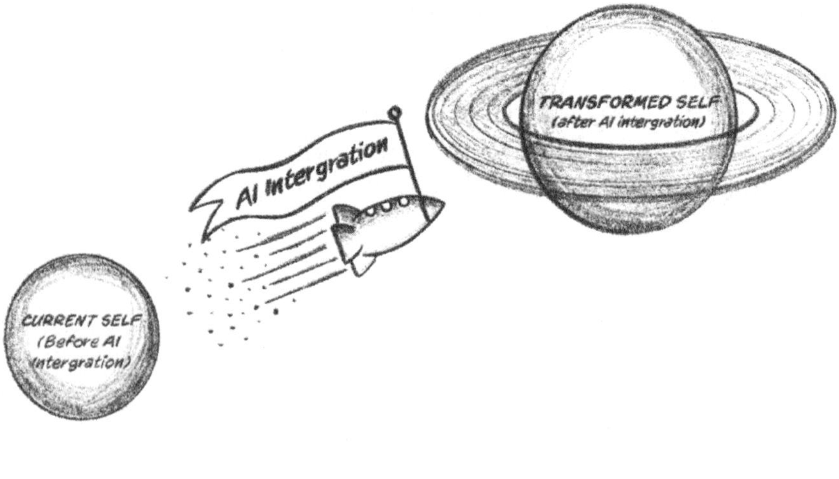

Envisioning the Transformed Self: Post-AI Integration

Now would be a great time to start thinking about what you want from your AI flywheel; what are the most important benefits to you? As an exercise, take a piece of paper and spend some time with your current self. Write down two or three parts of your life that you want to improve. They don't even have to be work-related; maybe one of your biggest frustrations right now is that you feel disconnected from your partner or that you don't have enough money to buy the new Xbox that's about to come out. Now, take a little more time and dream about your future transformed self, the self for which these visions are a reality. Finally, ask yourself: how can integrating AI into my workflow help me reach my ideal, transformed self?

This might be a straightforward exercise for you; you may have a list of things you'd like to see change! However, if you're struggling with this exercise, here's a list of some ways that you could transform yourself through AI integration:

- **You could be more productive at work.** We've already seen several studies demonstrating how workers have improved productivity by using AI, but just in case you're still unconvinced, here are three more. In Study One, customer service agents who used AI reported fielding 14% more calls per hour.[25] In Study Two, business professionals who used AI to help produce documents could draft 59% more documents per hour.[25] In Study Three, programmers who used AI to help create code reported an incredible 126% increase in projects completed per week.[25] Your transformed self will almost certainly operate at a similarly high level of productivity.
- **You could eliminate burnout from overwork.** Being more productive doesn't have to mean working longer hours. Your transformed self will show up to work inspired because you will get to spend the day doing the parts of your job that you love, and you will spend less time working overall because you won't have to worry about doing the repetitive parts of your job anymore.

- **You could be a confident, insightful decision-maker.** For example, say you work in consulting for universities and are trying to help a university decide whether to build a new performing arts centre. That's a substantial financial investment for the university, and your current self may be losing sleep over whether you'll get the decision right. However, in a similar situation, your transformed self will be sleeping easy, knowing that you have a series of AI-interpreted data on potential student enrollment impact, possible unforeseen construction costs, and prospective revenue figures that will help you make an informed decision.

- **You could be an asset to your employer.** As a productive employee who regularly makes intelligent decisions, it will be impossible for your employer to ignore you. Your transformed self will receive all kinds of opportunities from your employer to work on new and exciting projects or receive promotions. Alternatively, if you are considering a career change, acquiring AI integration skills will give you a competitive edge in the job market. You will have the confidence to apply for positions in companies requiring AI knowledge, which is becoming increasingly common. With these skills, you can transform yourself and become an attractive candidate for employers seeking AI-savvy professionals.

- **You could have more financial stability.** That Xbox you wanted to buy could be a simple purchase, as increased recognition from your employer will lead to promotions that will likely come with raises or other financial incentives. With more disposable income, your transformed self will spend less time worrying about a purchase's impact on your bank account.

- **You could feel more confident in both your professional and personal life.** A confident outlook on yourself is one of those things that permeates every part of your life. Improved work performance and increased comfortability in your personal life will help your transformed self walk with the sense that you're doing a good job. This increased confidence will feed right back into that AI integration flywheel, and the cycle will start all over!

Summary of Chapter Learnings

AI is the future, but it is also the present. Companies from nearly every industry are starting to incorporate AI, actively influencing how employees do their jobs. Amazon Go Stores are freeing employees from needing to act as cashiers, enabling them to fill other roles that AI cannot yet take, like stocking shelves or dealing with individual customer issues. ChatGPT is improving productivity and work quality for consultants, especially those consultants who were previously underperforming.[24] Uber is using AI to help its drivers choose areas of their region where they are likely to get the most customers and navigate traffic to speed up drop-off times.[28] McDonald's is using AI to adjust menu advertising based on everything from historical ordering data to the weather, and employees are learning to adapt to shifting consumer preferences.[29]

In today's fast-paced business world, companies rapidly adopt AI to stay ahead of the competition. As an employee, embracing this change and incorporating AI into your workflow is crucial. By doing so, you'll not only enhance your productivity but also stay relevant in the job market. If you're nervous about how much effort it will take or whether it will even work, think about the flywheel analogy. At first, it will take plenty of time and effort as you experiment with different ways to use AI. However, once you start seeing the returns on your investment in increased productivity, job opportunities, and confidence, you will be prepared to feed more AI integration into your flywheel. In time, working with AI will feel natural, and you will be a more confident, successful employee.

Becoming a more confident, successful employee is a fundamental part of the vision of the transformed self. Right now, you might have changes that you want to see in your life, whether professional or personal. By bringing AI into your workflow, you can start to effect the changes you want. Visualise your future self, a self that has high job satisfaction, plenty of job opportunities, a strong work-life balance, and confidence that carries through to every part of your life. AI integration can help you make that vision for your future self a reality.

6

MASTERING CHATGPT: UNLEASHING YOUR PRODUCTIVITY

There are an increasing number of AI tools available today, but one in particular has dominated the headlines: ChatGPT. The chatbot from the AI research organisation OpenAI debuted in November 2022, and its rise has been meteoric. But if you've never used a chatbot before, it might be hard to visualise what a chatbot is, what it does, and why ChatGPT has dominated the recent discourse on AI in a way that no other software has.

A chatbot is a software application or web interface designed to mimic human conversations through voice or text-based commands.[31] Simple chatbots have existed for decades, with a well-known early example being the ELIZA chatbot. ELIZA was first introduced in 1966.[33] It was designed with a basic set of rules embedded into its programming.[33] Whenever a user entered a prompt, ELIZA would scan the input for keywords, phrases, or patterns and generate a response accordingly based on what it found in the prompt.[33] Chatbots have come a long way since ELIZA, and today, they use deep learning and natural language processing to produce dynamic responses.

But even among contemporary chatbots, ChatGPT is special. Many chatbots are task-specific, meaning they can handle tasks like customer service responses or market research but are limited to functioning within that task category. However, ChatGPT is a highly versatile chatbot capable of engaging in conversations on various subjects, from simple maths equations to story writing to highly technical discussions on biochemical research. To build such a diverse chatbot, OpenAI developed an artificial intelligence framework called a generative pre-trained transformer, or a GPT.[32] The GPT employs a deep learning architecture that takes in information and learns by analysing the relationships between the information, like how humans read stories.[32]

When we read a story, we don't just read the individual words; we derive meaning from combinations of words ordered together to create something coherent. ChatGPT was trained on large datasets of uncategorised text, enabling it to analyse word relationships and patterns.[32]

ChatGPT's strength lies in its ability to recognise these patterns. Unlike simply memorising information, the AI can analyse a prompt it has never encountered before and generate an appropriate response by using its pattern recognition learning.[32] This pattern recognition capacity, in combination with access to near-unlimited information through the internet, gives ChatGPT its widespread appeal. Because it does not require specific prompts, any user can ask ChatGPT a question, and even if their grammar is awkward or the question is vague, ChatGPT's pattern recognition capabilities will help it reproduce a compelling answer. ChatGPT is also impressive because its language model is remarkably natural. Through analysing massive quantities of text, it has learned how to give responses using anything from super simple to highly technical language.

Thanks to its unique combination of power and versatility, ChatGPT has become the central node on which public discourse on AI has been based, and the rate at which it has been adopted for mass use has been nothing short of historic. Take a look at the following chart that presents the time for critical mass adoption of various world-changing technologies:

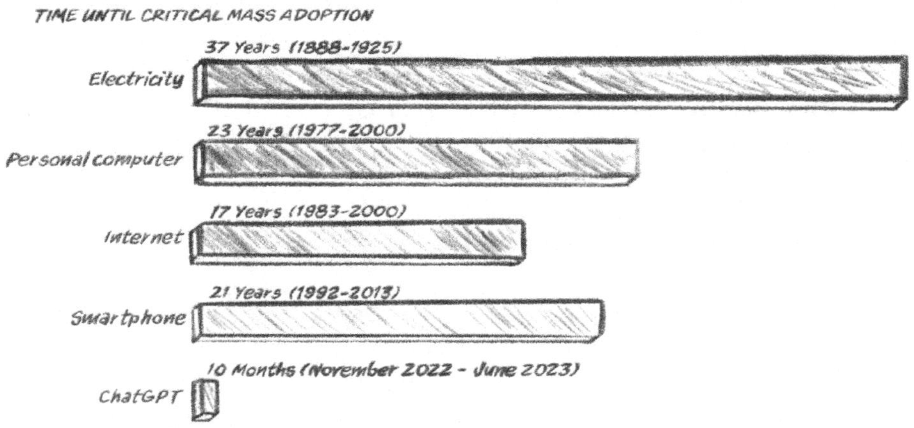

(Chart of critical mass adoption of world-changing technologies)

Note that critical mass adoption is defined as regular use by more than 50% of the surveyed population.[30]

In just ten months, ChatGPT achieved the kind of mass permeation that took the smartphone twenty-one years! As the standard-bearer for the AI revolution, ChatGPT's rapid adoption means that AI technology has reached critical mass adoption, putting it among the company of some of the world's most significant technological advancements.

Setting Up ChatGPT for Professional and Personal Use

OpenAI made ChatGPT extremely easy to access. To start using ChatGPT for personal requests, all you need to do is create a free account at their website: chat.openai.com.[31] You can sign in directly with a Google or Microsoft account or sign up with an email address.[31] From there, it's as simple as typing your request into ChatGPT's message bar. This is ChatGPT's most basic level of functionality, although even at its most basic, the software can handle many different requests.

Choosing a Subscription vs. Using the Free Access

ChatGPT started as an entirely free software, but as its

popularity exploded, OpenAI migrated the software to a freemium model.[31] At the time of writing, two plans are recommended for individual use and two for businesses. Note that OpenAI advertises its prices in USD, so the exact pricing within your locality may vary.

The individual plans are:

- **Free ($0/month).** ChatGPT recommends this plan for people who are new to the software and for everyday use.[34] The free plan comes with access to the older ChatGPT-3.[5] model, unlimited interactions, and access via either the website or the app on iOS or Android.[34]

- **Plus ($20/month).** ChatGPT recommends this plan for people who have been using the software for a while and want to expand their productivity.[34] The Plus program gives you access to (at the time of writing) the most updated version of the software, ChatGPT-4o, as well as access to other OpenAI tools like Dall·E, browsing, and advanced data analysis.[34] Alternatively, if you need more refined functionality in your chatbot, you can search through OpenAI's catalogue of GPT systems, or you can create your own. Note that this plan caps how many interactions you can have with ChatGPT and the other software.[34]

The business-level plans are:

- **Team ($25/month).** This plan provides all the individual benefits from the Plus membership with additional capacity for sharing GPTs between employees, higher message caps, and an admin console for workspace management.[34]

- **Enterprise (various/month).** Enterprise provides unlimited access to the entire suite of software, as well as additional admin controls and priority access to support.[34] The price of this plan is negotiated directly between OpenAI and the company.

You may also find that you only need access to some of the software provided with these plans; for example, you may need to do lots of image generation, but language models are less important for you. OpenAI also offers per-use pricing for access to individual software. For more information on the available software and prices, visit OpenAI's pricing page: openai.com/pricing.

Basic Interactions and Commands

Once you've selected your account and pricing plan, you're ready to start! The easiest way to learn ChatGPT is to simply jump in. Pick a few topics on which you want to test the software, evaluate whether the responses suit your needs, and adjust your requests accordingly.

There are all kinds of interactions you can have with ChatGPT, but here's a list of the most basic ones:

- **Ask Questions:** You can ask ChatGPT for information, like "What is the capital of Japan?" or "What time is it in Luxembourg if it is eight o'clock in London?"
- **Get Recommendations:** ChatGPT can offer recommendations based on your input. For example, suppose you ask ChatGPT for a one-day itinerary for sightseeing in Paris and specify that it should emphasise indoor activities but exclude the Louvre and the Eiffel Tower. In that case, an itinerary will be suggested based on these specifications.
- **Seek Information:** If you need to learn something new very quickly, ChatGPT can be an excellent resource. For example, if you are trying to build a connection with a potential client and you know they are interested in classic French films, you could have ChatGPT give you a crash course in French cinema by typing something like, "Describe in simple terms the history of classic French films and give me five examples."

- **Generate Text:** You can ask ChatGPT to generate text based on a prompt. For example, you could ask it to write an email to your coworkers announcing the retirement of a colleague, being sure to praise their long history with the company and their dedication to producing high-quality work.

Again, part of the trick with ChatGPT is creating prompts to generate the response you want. In most cases, specificity is your friend. For example, if you feel like ChatGPT is giving you information that's too complex, try specifically asking it to explain the material "in simple terms."[31]

As you're starting with ChatGPT, it may be hard to visualise what you can ask the software to do or how to phrase your requests. If you're having trouble getting started, check out the back of the book, where you'll find a list of 100 prompt ideas. These prompts run the gamut from meeting scheduling ("Can you assist in finding a suitable time for a team meeting?") to competitive analysis ("Can you help gather intel on our competitors and their strategies?"). Each prompt also comes with suggested information to provide ChatGPT so it can produce the best possible result (meeting duration, individual availability, and preferences for meeting scheduling; industry and key competitors for competitive analysis).

Maximising Output: Effective Prompting Strategies with ChatGPT

In the constantly evolving world of artificial intelligence, it is essential for professionals who want to take advantage of technology for improved productivity to learn how to interact effectively with AI models such as ChatGPT. This section focuses on the techniques of creating prompts that generate accurate and useful outputs. Whether you're writing reports, producing creative content, or analysing complex data sets, learning to communicate your requirements to ChatGPT clearly and efficiently can revolutionise your workflow.

By using targeted strategies in prompt engineering, you can unleash the full potential of AI to meet your professional demands. From the importance of precision and structure to the effectiveness of iterative dialogue, these guidelines will provide you with the necessary tools to engage with ChatGPT in a way that brings out the best of artificial intelligence.

1. Be Specific and Detailed

- **Precision in Requests:** Clearly articulate what you're asking for. Specificity helps ChatGPT understand the exact nature of your request, leading to more accurate and relevant responses.[32]
- **Detail Level:** Include essential details that can affect the response.[32] For example, if asking for a report, specify the format, length, and key points that need to be covered.

2. Structured Prompts

- **Logical Structure:** Present your request in a logical, easy-to-follow manner. If there are multiple parts to your query, enumerate them or break them down into separate prompts.[32]
- **Bullet Points:** Use bullet points for complex requests to delineate different aspects you're asking about.[32]

3. Context is Key

- **Provide Background Information:** Giving context helps ChatGPT tailor responses to your specific situation.[32] However, balance is crucial; too much can be overwhelming, and too little might result in generic responses.
- **Update the Model:** If the task is ongoing, update ChatGPT with the latest developments or information from previous interactions to keep the context current.[32]

4. Iterative Dialogue

- **Refine Through Conversation:** Treat the interaction as a conversation where you can refine, clarify, or expand upon your requests based on the responses received.[32]
- **Clarify and Correct:** If the output isn't what you expected, use follow-up prompts to guide ChatGPT in the right direction, specifying what needs to be changed or improved.

5. Utilise Examples

- **Show, Don't Just Tell:** When possible, provide examples of what you're looking for, such as a template, structure, or content style. This can significantly improve the relevance of the output.[32]
- **Positive and Negative Examples:** If applicable, showing examples of what you do and do not want can be incredibly instructive.

6. Command and Control Instructions

- **Direct Commands:** Use command-like language for tasks (e.g., "Write a summary," "Generate a list," "Explain the concept"). This straightforward, direct approach is often more effective.
- **Adjust Complexity:** Tailor the complexity of the language and concepts based on your needs or the content's intended audience.[32]

7. Feedback Loops

- **Iterate Based on Responses:** Use the feedback from each response to refine your prompts further. This iterative loop can significantly enhance the quality of outcomes over time.[32]

8. Understanding Limitations and Creativity

- **Work Around Gaps:** If you hit the model's knowledge or capability limits, rephrase or refocus your prompts to approach the problem differently.
- **Creative Problem-Solving:** Getting the best output sometimes requires thinking creatively about how to phrase your request or break down your task.

To implement the strategies effectively, it is important to practice and experiment to find what works best for each unique situation. Mastering the art of prompt engineering is not just about getting the correct answers from ChatGPT; it is about creating a collaborative dialogue where human creativity and AI capabilities can work together.

The strategies outlined in this section provide a foundational approach to harnessing the power of AI and are tailored to the dynamic needs of today's workplace. By being specific, structured, and strategic in your interactions, you can guide ChatGPT to produce outputs that meet your immediate needs and inspire new ways of thinking and working.

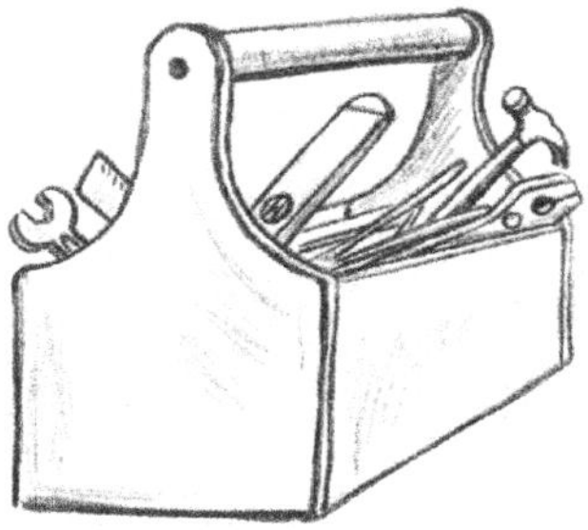

Adding to Your AI Toolkit: Complementary Tools and Resources

As powerful as ChatGPT is, its versatility means it might be less effective for high-level tasks or tasks requiring generating content other than text. In Chapter 3, we have a brief introduction to some of the software available to help you fulfil these needs. Here, we're going to dive a little deeper into those tools. For each software, you'll get the user difficulty level, 2024 price (in USD), key features, and an assessment of productivity gains you will achieve by using the software. Note that some of this information (especially the price) is subject to change, so it's best to check with the company for the most up-to-date information.

Jasper AI

- **User Difficulty Level:** Beginner
- **Pricing:** The basic plan starts at $29/month, and more advanced plans offer additional features and higher usage limits.
- **Key Features and Uses:** An AI-powered content generation tool that helps create written content efficiently.
- **Productivity Gains:** Enhances content creation speed and quality; ideal for marketing, blogging, and email communication.

Copy.ai

- **User Difficulty Level:** Beginner
- **Pricing:** A free trial is available; Pro plans start at $35/month and provide unlimited runs and access to premium features.
- **Key Features and Uses:** Generates creative copy for ads, emails, websites, and more, using AI to streamline the copywriting process.
- **Productivity Gains:** Accelerates the development of compelling marketing and communication materials.

Synthesia

- **User Difficulty Level:** Beginner to Intermediate
- Pricing: Subscription plans start from $30/month, which includes the creation of up to ten minutes of video per month.
- **Key Features and Uses:** Creates AI-generated videos from text inputs, eliminating the need for traditional video production resources.
- **Productivity Gains:** Offers a cost-effective and efficient method for producing educational, marketing, and informational videos.

Descript

- **User Difficulty Level:** Beginner to Intermediate
- **Pricing:** A free version is available; the Pro version starts at $12/month and offers advanced editing features and more cloud storage.
- **Key Features and Uses:** Provides comprehensive audio and video editing tools with auto transcription and overdubbing features.
- **Productivity Gains:** Simplifies multimedia editing, making it quicker and more accessible, especially for content creators.

Canva's Magic Write

- **User Difficulty Level:** Beginner
- **Pricing:** A free tier is available, and a Pro subscription, featuring more advanced design and collaboration tools, starts at $12.99/month.
- **Key Features and Uses:** It generates text for various design

purposes and is seamlessly integrated with Canva's design platform.

- **Productivity Gains:** Streamlines the creation of visually appealing designs with relevant and engaging content.

Runway ML

- **User Difficulty Level:** Intermediate to Advanced
- **Pricing:** A free tier is available, and paid plans with additional features and computing power start at $15/month.
- **Key Features and Uses:** Offers advanced AI tools for creative projects, including video editing, music generation, and more.
- **Productivity Gains:** Enables creative professionals to explore and create with AI technologies, enhancing innovation and productivity.

Tableau CRM

- **User Difficulty Level:** Intermediate
- **Pricing:** The Tableau CRM Growth plan starts at $75/user/month, with costs varying based on features and the scale of deployment.
- **Key Features and Uses:** Integrates artificial intelligence and machine learning for data analysis and visualisation, facilitating insights from complex datasets.
- **Productivity Gains:** Streamlines data analysis and interpretation, enabling data-driven decision-making with less effort and more accuracy.

Gemini (Formerly Known as Google Bard)

- **User Difficulty Level:** Beginner

- **Pricing:** A free tier is available; a $19.99/month plan with extra features is available, as is a business-level plan (for business-level plan pricing, contact the company).
- **Key Features and Uses:** Similar to ChatGPT, it can generate conversational outputs, provide explanations, offer summaries, and create content based on the prompts it receives.
- **Productivity Gains:** Enhances efficiency in research, content creation, and idea generation processes.

QuillBot

- **User Difficulty Level:** Beginner
- **Pricing:** Free tier available; Premium subscriptions, which unlock full features like expanded paraphrase modes and higher character limits, start at $8.33/month when billed annually.
- **Key Features and Uses:** This tool paraphrases and summarises text and provides rephrasing suggestions to improve clarity, avoid plagiarism, and tailor the style to a specific audience.
- **Productivity Gains:** Streamlines the writing process by offering quick alternatives to express the same ideas differently, enhancing the originality and readability of texts.

Grammarly

- **User Difficulty Level:** Beginner
- **Pricing:** A free tier is available. Premium plans, which provide more comprehensive writing feedback and features, start at $12/month when billed annually. Business plans are also available, starting at $15/month when billed annually.

- **Key Features and Uses:** Provides spelling, grammar, and punctuation corrections and stylistic creative writing suggestions.
- **Productivity Gains:** Beyond improving writing efficiency and quality, Grammarly aids in brainstorming, helping users overcome writer's block and enhance creativity in content development.

Summary of Chapter Learnings

In the past two years, ChatGPT has grown from a relative unknown in the burgeoning AI field to a global phenomenon. OpenAI debuted the chatbot in late 2022, and upon launch, it was widely praised for its ability to mimic natural language and its versatility. ChatGPT is capable of such diversity because it is built on a learning program that memorises information and learns patterns in text. This pattern learning ability enables ChatGPT to take in a prompt that it has never seen before, determine what the prompt writer is likely asking based on its knowledge of word patterns and produce an effective response.

Because ChatGPT is so versatile, it has been widely used in personal and professional settings. If you're not already using ChatGPT, now's a great time to start! OpenAI provides basic (but still powerful) access to ChatGPT for free; all you have to do is create an account and start talking to the chatbot. As you get used to communicating with ChatGPT, you will learn how to formulate your prompts to maximise the benefit of the responses you receive. In particular, the more information you can give the chatbot, the more nuanced its response will be. Once you master the basics of ChatGPT, if you find that you want more flexibility or access to new tools, you can look into OpenAI's subscriptions, which will include the latest version of ChatGPT and access to other OpenAI software like the image generation software Dall·E.

Once you get your AI flywheel spinning, you will encounter more and more situations that could be simplified with AI, and some of these situations may be suited to a more specific AI software. Examples of powerful but slightly more targeted AI include Jasper AI (content generation), Synesthesia (AI-generated videos), and Tableau CRM (data analysis and interpretation). If you can visualise a scenario where AI would make your life easier, there's most likely software out there that could make it happen!

7

AI-DRIVEN DECISION MAKING AND PROBLEM SOLVING

We are facing more decisions than ever before. Just think about all the choices you've had to make in the past 24 hours. You might have gone to work and had to decide which task to complete first. Once you decide on a task, you have to determine the best way to approach it. Perhaps you had to figure out what the outcome of your work on that task would be and then decide on your next task. After doing that all day at work, you come home and have to decide what to have for dinner, which TV show to watch, what time to go to bed, and the list goes on.

In addition to the increasing number of decisions we must make, we are also being bombarded with more data to influence us in those decisions. The resulting combination of more decisions and more data is leaving people overwhelmed by the decision-making process. In a recent survey, seventy-four per cent of respondents said the number of decisions they've had to make had increased tenfold over the past three years, and seventy-eight per cent said their stress over decision-making is partly attributable to information overload.[36] Fifty-nine per cent of respondents reported facing daily decision-making dilemmas where they were unsure how to proceed.[36]

Luckily, we live in the AI era, which means we have the tools to take some of this decision-making burden off ourselves. AI has a diverse future in decision-making assistance. For example, logistics companies have started deploying AI to track the origins of their raw materials and make preventative action decisions if they find that the materials are being produced in an unsustainable way. Other companies have started using AI as a decision-making trainer, generating scenarios for trainees to practice decision-making skills in virtual environments. However, one of the most powerful deployments of AI today is in data analytics.

AI Tools and Techniques for Data-Driven Decision Making

The age of information is here to stay, and as a result, we need to develop strategies for making decisions in a world where we are often presented with an abundance of data. Fortunately, AI is highly proficient in handling and analysing data and can enhance data analysis in several ways, primarily by enhancing efficiency, accuracy, and scalability. Here are some of the ways that AI can improve data analysis:

- **Automated data processing.** Artificial Intelligence (AI) algorithms have the capability to automate routine data analysis tasks such as data normalisation, cleaning, and transformation.[37] By automating these tasks, employees can save significant time and effort, which can then be redirected towards more complex tasks that require human expertise.
- **Pattern recognition and anomaly detection.** Once data has been entered into an algorithm, AI can learn to recognise complex patterns in the datasets that may not be easily noticeable to humans.[37] This ability to identify patterns can then be used to detect anomalies in the dataset.[37] Anomaly detection can be beneficial for detecting fraud or identifying points where sales are not meeting expectations.[37]
- **Predictive analytics.** When AI learns to recognise patterns, it can use those patterns to forecast future outcomes.[37] This can be useful for various predictive tasks, including sales forecasting, demand prediction, and risk assessment.[37]
- **Natural language processing.** NLP enables AI to analyse text-based data like social media posts and customer reviews.[35] NLP-powered tools can inform decision-makers about customer feedback, market trends, and public sentiment.[35]
- **Image and video analysis.** AI can use image recognition

to classify objects and track faces, while video analysis can identify activities and behaviours.[37]

In addition to its diverse functionality, AI derives utility from its scalability. Cloud-based AI platforms have access to massive computing power, enabling them to analyse huge datasets quickly and efficiently without significant upfront investment in infrastructure.

Implementing AI for Strategic Advantage

Taking a structured approach to implementing AI into your decision-making workflow is best. This will give you a functional, repeatable methodology and help you precisely define the parts of your approach that need restructuring as you add more AI integration. Use the following step-by-step approach as inspiration to design a process that's tailored to your needs:

1. **Define objectives and use cases.** Establishing clear expectations for what you aim to accomplish by utilising AI and how you will incorporate it is important.[37] For instance, you may leverage AI to determine when to place resupply orders, and you may implement a predictive analytics model as your use case. As you define your objectives and use cases, consider any potential ethical quandaries.[37] Ask whether your AI use case follows regulations and protocols and whether it's being implemented in a way that reinforces biased behaviour.

2. **Collect your data.** This step is self-explanatory. Prepare the data you want AI to analyse, whether it's quantitative datasets, images, or text-based consumer reviews.

3. **Select the appropriate AI technologies**, or develop your own. Refer to the software lists provided in Chapters 3 and 7 for recommendations on software that may suit your specific use case. If you are developing your own module, it is crucial to adopt a human-in-the-loop approach. This means that the AI should be able to explain its results in a manner that humans can comprehend and that humans

will have the final say on all decisions.

4. Validate the performance of the AI through evaluation and testing. It is essential to conduct regular audits of your AI to ensure it meets your objectives. If the results are not as expected, don't hesitate to retrain your software or switch to a new one. Suppose you are working with other employees on the same software. In that case, it is a good idea to establish a feedback mechanism so that everyone can contribute to improving your AI's performance.

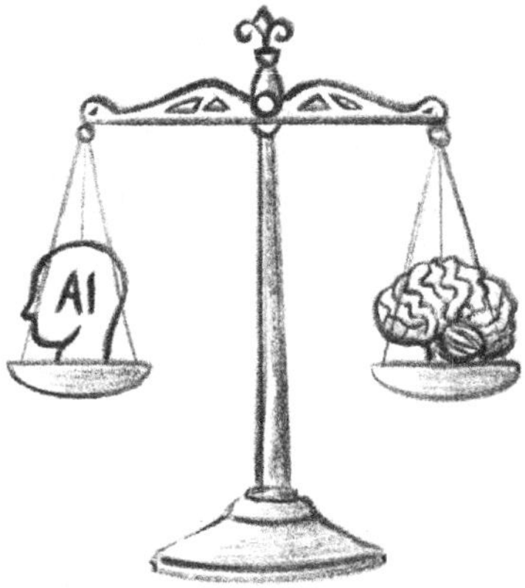

Balancing Human Judgement with AI Insights

The AI Employment Paradox theory has raised concerns that AI could replace human workers by mimicking human thought processes. But as we've discussed, that's a highly unlikely outcome. In fact, the most likely outcome of AI's rise is a complementary relationship between AI and human intelligence.

AI and humans have complementary strengths, not overlapping ones. To describe this mutualistic relationship, we can say that there are three ways in which human capabilities assist machines and three ways that machines assist humans.

First, humans can teach. AI algorithms are powerful but useless if they're not told what to do or how to do it. Humans can selectively feed datasets to AI based on what they think will be useful for helping AI learn to do its job.[39] Additionally, AI has to be trained on how to interact with humans. Humans have to teach AI how to produce results that are meaningful and interpretable by a human.[39] In cases where AI is interacting with the general public, humans have had to train AI to showcase a certain human-like personality. For example, Microsoft's Cortana was trained to be confident, caring, and helpful but not bossy, whereas Apple's Siri was trained to have just a hint of sassiness.[39] For the general public, AI is much more enjoyable to interact with when it has a personality.

Second, humans can explain. AI can produce a ream of fascinating data analysis, but all of that analysis is meaningless if it's not interpreted. Humans have the capacity for critical thinking, meaning we can take the results of an analysis and determine what it means in terms of a pending decision.[39] Additionally, AI needs human subject matter experts to explain analysis results to non-experts. Even though we've started factoring algorithms into weighty decisions like legal or medical matters, a human touch is still essential for disseminating information in a situation that could be emotionally charged. In fact, The European Union has officially recognised citizens' rights to receive a human interpretation for any AI-generated response to credit card or mortgage decisions.[39]

Third, humans can sustain. Ultimately, machines are human-built, and machines suffer breakdowns. Until machines learn how to self-repair or self-correct, we will need humans to maintain them.[39] This could mean either physical repairs, like replacing broken hard drives, or software repairs, like drawing up new training strategies after a self-driving car is responsible for a traffic accident.[39]

The first way that machines can assist humans is by amplifying. Artificial intelligence can enhance our decision-making by providing the correct information at the right time.[39] AI can impact both analytical and creative decisions. If a fashion designer created a new shirt, they could ask the AI to calculate based on size, material availability, and potential consumer market. This would free the designer up to focus on parts of the iterative process that require a human touch, like aesthetics.

Second, machines can interact with employees and facilitate employee-employee interactions. For example, transcription software can record a meeting and interpret the highlights, which could then be sent to employees as a reminder of the meeting's takeaway messages or as a summary for employees unable to attend.[39]

Third, machines can embody. Because they are built and trained by humans, AI algorithms naturally take on specific human-like characteristics.[39] Machine designers are using this to their advantage, building robots to do jobs that embody the essential work that humans can do but do it better, faster, and safer. We often talk about AI mimicking human intelligence, but it can also augment human strength. For example, robotic arms equipped with AI technology might handle the parts of a task that require heavy lifting.[39] In contrast, a complementary human employee handles the parts that require dexterity and human judgement.[39]

Challenges in Integrating AI with Human Decision-Making

As we've discussed, AI is revolutionary, but it is not perfect, and there is an inevitable learning curve associated with bringing it into the workplace. One of these challenges is that, for many people, AI is still an emerging technology. Although we've been interacting with AI for years, for many of us, our conscious interaction with AI has been limited to simple interactions with chatbots or voice assistants like Siri and Alexa. The leap to using AI for impactful decisions might be a terrifying one for you. This may be because you don't understand how the algorithms work, don't trust the AI to be truly unbiased and accurate, or are nervous that the learning curve will be too steep.

These fears are natural, but they are worth overcoming because AI offers so much potential to revolutionise the decision-making process. If you are the decision-maker on a project, dive headfirst into using AI and intimately familiarise yourself with its functionality. That way, when it comes time to make an AI-informed decision, you will be more comfortable because you know how the AI generates its information. If you are the one using the AI but others are making the decisions, learn to explain the rationale behind the AI's conclusions and recommendations, and come equipped with case studies and real-world examples (like those we've discussed in this book) that demonstrate AI's potential.

If you are worried about bias, keep a close eye on your training data to ensure the AI is not being trained on biased information. As the AI produces results, trust your critical thinking ability to sniff out biases and then correct the AI through additional training.

You've heard this refrain several times now, but it's so important that it bears repeating: the best way to get comfortable with AI is to start using it. Suppose you're nervous about bringing AI into your decision-making, whether due to a lack of understanding, concerns about bias, or fear of the learning curve; the easiest way to assuage those nerves is to start practising. You can start small, have an AI algorithm analyse your tasks to decide which you should work on first, or ask it to analyse a small decision that is part of a larger project. These small steps will get your AI flywheel spinning and help you grow your confidence.

Summary of Chapter Learnings

Making decisions nowadays has become increasingly challenging. On the one hand, we have to make more decisions than ever before, and on the other hand, we are constantly exposed to a vast amount of information that is supposed to help us make those decisions. However, the abundance of information often leads to information overload, making it challenging to sort through it all making decision-making even more difficult.

Although we may not be able to reduce the number of decisions we have to make, we can make those decisions much easier thanks to AI-assisted analytics. If your decisions require a lot of data, AI can automatically clean up and sort your datasets. AI can also be trained to recognise patterns in datasets, and thereafter, it can analyse new datasets to either reinforce patterns for trend prediction or discern anomalies. Natural language processing and image and video analysis allow AI to turn qualitative sources into quantitative information that employees can use to make decisions based on customer feedback or behaviour analysis.

Incorporating AI into your decision-making process requires a structured approach similar to the scientific method. Firstly, the objectives and use cases for AI integration must be identified, and both AI and human roles must be well-defined to highlight complementary work. Secondly, train the AI to produce the necessary information and then incorporate it into your decision-making process. As you gain insight into the AI's work and the results it produces, evaluate whether the integration has achieved your established objectives. If it has, start integrating AI into a new area of your decision-making process. But, if it has not, refine the AI's training, switch to new software, or re-evaluate your objectives until you get the desired results.

Far from diametric opposites, AI and humans have complementary skill sets. Humans are responsible for training AI to align with their objectives, interpreting the output AI produces, explaining it to non-experts, and maintaining the AI systems. AI enhances human capabilities, improves efficiency, and fosters interactions between humans while embodying human-like traits.

The challenges humans face as they integrate AI into decision-making reflect some of the general concerns surrounding AI. You may be hesitant to rely on AI to make significant decisions because you lack a clear understanding of how it arrives at its conclusions. You could be concerned about the possibility of AI being influenced by bias or prone to errors. Or, you might be worried that learning new skills required to work with AI might be too challenging. However, these fears can be mitigated by simply familiarising yourself with AI. If you intentionally dedicate time and energy to the practice, you will soon find it becomes a natural extension of your decision-making process.

8

AI AND YOUR WALLET REALISTIC STRATEGIES FOR EARNING MORE

Disclaimer: *The information provided in this chapter is intended for educational and informational purposes only and does not constitute financial advice. The tools, strategies, and examples discussed are for general reference and do not guarantee specific investment results. Readers are encouraged to conduct their own research and consult with a qualified financial advisor before making any investment decisions.*

Negotiating salaries and promotions can be a nerve-wracking experience. You may be concerned that asking for a raise or discussing the possibility of a promotion could be considered unreasonable or negatively impact your professional image. However, the reality is that salary and role discussions are part of being a professional, and the rise of AI will not change that fact. However, AI will impact how we talk to our superiors about our careers. Therefore, it is important to understand how AI impacts the job market and how developing certain skill sets can increase bargaining power.

As we've discussed, jobs from nearly every sector will be impacted by AI; most of these jobs will be redefined to complement the work that AI can do, while a few may disappear. We've also seen several examples of job roles and responsibilities being redefined thanks to AI integration. Chatbots have emerged as the front-line workers of the customer service industry; finance workers increasingly need data analysis and machine learning skills to engage in algorithm-based trading and conduct risk assessments; business owners are focusing more on personalised marketing campaigns.

However, note that productivity is a common thread connecting these shifting job descriptions. AI integration enables employees to interact with more clients, make more decisions, and increase creative output. In today's world, where there is a lot of competition for potential client's attention, producing fast, efficient, and high-quality results is crucial. If you, as an employee, can increase your productivity without overworking yourself, you will have more leverage when it comes time to negotiate a raise or a promotion.

So, what skills do you need to succeed in an AI-powered world, and how will those skills help you compete in the job market?

The most concrete skill that you will need to acquire is tech-savviness. At the most basic level, it is essential to have a fundamental understanding of computers—how to develop a good file storage system, access the internet and cloud-based resources, find information through Google searches or by asking ChatGPT, etc. If you have solid computer literacy, the next step is understanding how machine-learning algorithms work and how to manipulate them. You don't have to become an expert, but a basic understanding of topics like regression, classification, clustering, and neural networks will help you understand how AI works and how you can manipulate it. As you learn to operate algorithms, developing some basic programming skills will be helpful. Becoming proficient in basic programming languages like Python or R will help you manipulate and debug the code used for machine learning training.

Apart from concrete tech knowledge, most of the skills

that will be valued in an AI-powered world are soft skills. For example, employees must show strong problem-solving, critical thinking, and innovation skills. AI integration is the product of dreamers, pioneers who see an opportunity to increase productivity and take advantage of it. An AI-savvy employee will identify new opportunities for AI integration, successfully interpret the results provided by the AI they are already using, and troubleshoot when the AI is not functioning as expected.[45]

Even with its impressive abilities, AI still cannot replicate the richness of human interaction. Therefore, an AI-savvy employee will need people skills and a high degree of empathy. In the case of customer service representatives, chatbots may be the new front lines of customer interaction.

However, a human touch is still necessary if the customer is having a more complicated or sensitive issue. Jobs that have historically been human-facing will likely become even more so; as AI assumes the parts of the job that do not necessitate human interaction, employees will spend more time interacting with the people who define their jobs.

Finally, employees will need to become lifelong learners. This means engaging with new technologies as they become available and learning to be adaptable.[45] An AI-savvy employee will take advantage of continuing education programs to stay informed about the latest AI advancements, and they will welcome shifting job responsibilities as an opportunity to grow.[45]

Developing an AI-savvy skillset is the best thing you can do to increase your bargaining power when it comes time to negotiate a raise or a promotion. And if you're worried that all this training and skill development could go to waste if your company doesn't recognise the power of AI in your area of expertise, fear not! As we've seen, three out of four employers now prioritise hiring AI-savvy employees, and those same companies are willing to pay more for such employees. Furthermore, it is not just the employees who build and maintain the algorithms that will see financial gains—while these employees stand to average 47% more income, sales and marketing employees stand to average 43% more income, 42% for finance, 41% for business operations; 37% for legal and compliance; and 35% for human resources.[41]

Sander van't Noordende, the CEO of talent company Randstad, concurs, saying that employers are "going to be paying [personnel] more to do a job that adds even more value."[46] He has also commented on the relationship between AI and productivity, saying, "AI increases productivity, helps you do a better and a faster job so that you can spend more time with your clients... [or] with your colleagues, venturing into new areas so it's freeing up time to do more high value things."[46]

When it finally comes time to negotiate, AI can empower you to confidently make requests. Use an AI-powered tool like Payscale to analyse thousands of job listings similar to the one you are negotiating for.[40] These tools can offer insight into fair salary ranges and the benefits you can ask for, like flexible hours or remote work options.[40]

Using AI for Job Searches and Identifying Skill Gaps

You may find that the best source of career advancement is outside your current company. If you think that job advancement could mean searching for a new employer, you will find yourself going through the job hunt process. As important as it is to showcase AI skills on your job applications, you can go a step further and utilise AI to assist in the job search process itself.

AI's Influence on Hiring

Before we discuss how to use AI to help with your job search, it's essential to understand how AI is impacting the hiring process on the other side. Human resources departments are increasingly utilising AI to automate the hiring process, tasking AI with sorting resumes and screening cover letters before handing a select group of applicants to a human reviewer.[47] Algorithms are often used to scan job applications for specific keywords.[47] If the applicant's application doesn't include these keywords, their application may be rejected. Once a candidate passes the initial screening and is invited for an interview, AI-powered chatbots might be used to conduct the first round of interviews.[47] During this stage, the chatbot will ask predetermined questions and collect real-time responses from the candidate.

Because AI has become widespread in human resource departments, it's important to recognise that you will likely encounter AI screening during the job search and application process. Therefore, you will need to adjust your approach to finding openings and preparing application materials accordingly.

Leveraging AI-Powered Job Search Platforms

Take advantage of job boards that use AI to match your profile with jobs that suit your skills and qualifications. A job board like Talentprise is an excellent example of this. Prospective employees complete a questionnaire about their job preferences, write a biography, and take an assessment to gauge their personality traits, values, and motivation via Talentprise.[40] With those data points in place, Talentprise's AI matches the prospective employee with relevant job postings.[40]

When using an AI-powered job search platform, filling out your profile as completely as possible is important. If you provide more information, the AI's algorithm will be better equipped to match you with relevant job openings.[40]

Using AI to Optimise Your Resume and Cover Letters

To increase your chances of moving forward in the hiring process, it's important to customise your job application by including specific keywords that the AI-powered human resource system will scan.[40] This will help the system identify your qualifications and skills, making your application more likely to be selected for the next stage of the hiring process. If writing a resume or cover letter is not your strong suit, you can use software like Kickresume or CoverDoc.ai.[40] With this software, you can input basic information about yourself and the job you're applying for, briefly describe your writing style, and connect your LinkedIn profile.[40] Then, the software will generate a resume or cover letter. Furthermore, Jobscan and Skillsyncer are software tools that can compare your resume and cover letter to a job description.[40] They assess the degree to which your documents reflect the job description and suggest keyword implementation.[40]

Identifying and Bridging Skill Gaps with AI

If you are concerned about gaps in your resume, an AI-powered skill assessment can identify strengths and weaknesses based on your career goals. Once you have identified these skill gaps, seek opportunities to fill them. Continuously pursue new learning opportunities; take continuing education courses if your company offers them, especially if they are related to studying AI. If you need to learn a new skill, seek out an AI-powered course that you can adjust to your skill gaps and preferred learning pace.

Financial Growth for the Everyday Professional

If you use your AI-enhanced productivity to negotiate a higher salary, you'll have more disposable income available. To turn that extra income into even more money, consider using it to build an investment portfolio. This can help you grow your wealth over time. These days, artificial intelligence is widely used in financial services because its ability to process large datasets gives it the power to predict future market trends.

Here are some of the ways that you can use AI to manage your investment portfolio:

- **Stock picking.** When it comes to investing, choosing the right stocks can be daunting, given the numerous data points associated with each stock. This is where AI-powered stock screeners come in handy. These tools can sort through vast amounts of data and give investors the information they need to make informed decisions.[44] AI-powered stock screeners use algorithms to analyse a company's financial data, including earnings, revenue, and debt-to-equity ratios.[44] They can also assess a company's valuation relative to its peers and provide insights into its future growth prospects.[44] Several powerful free stock screeners are available, including Morningstar, Finviz, YahooFinance, and ZACKS.[44] These tools offer a wealth of information on individual stocks and sector and industry trends.

- **Automated portfolio building.** If assessing individual stocks is too tedious for you, consider using an automated portfolio builder. Also called robo-advisors, automated portfolio builders can evaluate your investment goals, timeline, and risk tolerance to assemble a list of stocks that meet your needs.[44] Furthermore, automated portfolio builders can automatically swap stocks in and out of your portfolio if they find certain stocks no longer meet your set parameters.[44] Wealthfront and Betterment are solid choices for introductory automated portfolio builders.[44]

- **Trading and trade management.** Choosing when to buy or sell a stock can be challenging, as you can never be entirely sure how a stock will fare once you move on it. However, you can mitigate some of the associated risks by having AI handle the trading process. You can configure an AI-powered trade manager to execute stock purchases only when certain conditions are satisfied.[44] Additionally, you can automate an exit strategy based on either the specific stock's performance or the general market performance.[44]

- **Data interpretation and predictions.** We've discussed several sectors where AI contributes to data interpretation, and stock trading is another example. In a prime example of AI complementing human intelligence, you can set AI to analyse the market based on your personal philosophy of how the market operates.[44] For example, if you tend to believe that stocks go through repeatable cycles of growth and contraction, you can ask AI tools to identify these cycles and predict when the next one will occur.

- **Risk management.** In addition to setting automated conditions for an exit strategy, you can use AI to reduce over-exposure in your portfolio. For example, AI can determine whether you have overvalued a stock and put too much money into it or if your portfolio has too many stocks in the same market or industry.[44]

Building an investment portfolio is one of the best ways to ensure long-term financial stability, and the rise of AI investment tools has made it easier than ever to safely and efficiently manage an innovative, diverse portfolio.

Summary of Chapter Learnings

AI is shifting the nature of work across job sectors. Job descriptions in the fields of customer service and finance are changing as AI automation increasingly finds a foothold in the workplace. However, employees who embrace this technology will see tremendous productivity gains. As an employee's workflow becomes more AI-automated, they can get work done faster and more accurately. Any company loves a productive employee, so increased productivity will lead to more recognition from the company, thereby increasing the employee's negotiating power.

An employee needs to develop a certain skill set to fully take advantage of AI's productivity potential. A basic understanding of how AI works and how to manipulate it will be necessary, as will soft skills like critical thinking, problem-solving, empathy, people skills, an insatiable desire to learn, and adaptability.

If you start to feel like career advancement is leading you toward a company change, knowing how AI is used in human resource departments and how you can integrate AI into your job search will be helpful. Human resource departments are using AI algorithms to scan applications for keywords, and if your application is missing some of these keywords, your application may get rejected. Therefore, it's essential to tailor your application to each job. Use AI-powered software to help you match with jobs that suit your skills and experience and tailor your resume and cover letter to a given job description.

AI can also help you reach a state of financial security by managing an investment portfolio. You can use AI-powered software to research individual stocks or build a portfolio that meets your criteria. When assembling a solid investment portfolio, it is essential to manage risks. To that end, AI can manage an exit strategy if a stock stops performing, or it can analyse your portfolio for potential over-exposure risks.

Of course, AI should not be limited to personal use and individual employee productivity gains. In the next chapter, we'll examine ways that you can use AI to streamline workplace communication, manage projects, and optimise team assembly.

9

COMMUNICATING WITH AI: ENHANCING COLLABORATION

In the modern workplace, nobody works inside of a bubble. All employees have to communicate to some degree, whether it's with colleagues about the status of a project or with a boss about shifting job descriptions. Furthermore, there are an increasing number of mediums through which to communicate in the workplace. Email and face-to-face conversations have been the mainstays of workplace communication for a long time, but now there are many platforms to choose from, like Slack, Asana, and Google Workspace.

Workers are feeling the stress that comes with increased quantities of communication. A recent survey found that 38% of employees and managers think they are bombarded by excessive, often confusing messages, and in the fray, important information often goes unread.[48]

However, as with so many other facets of the workplace, AI can positively impact workplace communication. The impact of AI is already being felt in how employees communicate; more employees are leaning towards concise, direct communication styles, which are influenced by the precision required when working with AI. AI is also leading a push towards instant messaging and collaboration tools instead of emails and face-to-face meetings.

We've already touched on some ways that AI can complement interactions between employees: note-taking tools for meetings, language translation tools for companies with an international employee base, etc. But now, let's dive deeper into this important area of AI workplace integration.

Let's first examine some more ways that AI can enhance employee communication:

- **Focus group summaries and sentiment analysis.** Joel DiGrolamo, the vice president of research and data science at the International Coaching Federation, says, "We have discovered that using very specific prompts when working with generative AI can synthesise focus group and workshop transcripts to provide an alternative look at the information provided in those sessions."[48] In other words, if you and your colleagues meet for a focus group, a work session, or a planning meeting, you can then ask an AI algorithm to review the session and provide additional insights. Similarly, AI can assess employee feedback through sentiment analysis, which can help increase employee satisfaction and promote a more diverse, inclusive workplace. Having AI participate in focus groups and sentiment analysis is like having another (very insightful) employee at the table.

- **Creating presentations.** Robert Garcia, also of the International Coaching Foundation, claims that as many as 80% of conference presenters have used AI to lay the groundwork for their presentations.[48] However, this is not a disingenuous practice; using AI to create presentations that will eventually be given to colleagues or bosses can ensure factual accuracy and improve audience comprehension.[48]

- **Chatbots.** We've discussed chatbots at length, but there is a reason for that: they are extremely powerful and versatile. In the case of workplace communications, chatbots can facilitate information distribution, provide a first line of intake for employee questions, and create personalised training modules.[48]

- **Personalisation.** For a large company with employees working in multiple sectors, AI can place employees into communication groups based on job role, location, interests, etc. In this way, professionals can send targeted communications. For example, an algorithm can be provided with a list of information to be transmitted to employees. Based on this list, the algorithm can determine the relevance of the information to each employee and only send them the information they require.

AI-Powered Tools for Workplace Communication

Anyone looking to integrate AI into workplace communication will be met with many options; many of the most familiar names in the tech space have started incorporating AI tools into their already-established platforms. Here's a list of some communication tools you may already be familiar with and an explanation of how they incorporate AI into employee communication:

- **Slack.** The favourite communication tool of many cutting-edge companies uses AI-powered features like smart search, suggestions, and reminders to streamline communication and collaboration.
- **Microsoft Teams.** Teams comes with features like real-time translation, meeting transcriptions, and intelligent scheduling assistance.
- **Google Workspace.** Google users can access AI-powered tools in most of the company's software, including Gmail, Google Drive, and Google Calendar. Features include Smart Compose (for wording suggestions when typing emails), Smart Reply (automated reply generation based on conversational context), and automatic file organisation.
- **Zoom.** Suppose you work remotely but still need to communicate with colleagues or clients in real time. In that case, Zoom uses AI to make the meeting experience run smoothly, with features like background noise suppression and automated transcription.
- **Microsoft Outlook.** AI-powered features through Outlook include Focused Inbox (for automated email sorting), predictive email scheduling (for ensuring that emails are sent at a time when they are most likely to be seen), and suggested replies.
- **Grammarly.** A slightly different component of communication facilitation, Grammarly offers grammar, style, and tone suggestions to ensure that communications

are easily understood and of an appropriate tone for the intended audience.

- **Slido.** Slido may be a lesser-known software, but it fills an important role: it uses AI power to determine levels of audience engagement. Suppose you are giving a presentation at a conference or to a room of colleagues. In that case, Slido can gauge audience participation, generate meeting engagement scores, or conduct sentiment analysis to discern the audience's mood during different parts of your presentation. Slido integrates with a variety of platforms, including PowerPoint, Google Slides, and Zoom.

AI and Project Management

Project management is a subset of employee communication that often takes up a disproportionate amount of a manager's time. Like most employees, effective project managers are always looking for new ways to improve efficiency and promote collaboration among the employees they manage. AI integration is an easy way to achieve these goals.

Here are a few ways that AI can be integrated into project management:

- **Team assembly.** This is one of the most critical assignments given to a project manager; without a well-functioning team, a project will most likely never get off the ground. AI-powered tools can sift through resumes, past project history, and professional profiles to determine which employees are the best fit for different parts of the project.[51] AI can also conduct a fit analysis, processing information on personality traits and work styles to determine how well employees will work together. For example, suppose two employees are known to clash regularly. In that case, a project manager can add this information to the algorithm to ensure that they are either not working on the same project or not working closely together.

- **Accuracy and forecasting.** By utilising an array of AI-powered predictive analytics, a project manager can assess where bottlenecks in project progression may arise or examine prior employee output to determine the parts of a project to which various employees should be assigned.[49] Most of the concerns that project managers deal with involve project timelines and budgetary concerns. AI-powered algorithms are especially well-equipped to address these concerns so project managers can confidently make decisions.
- **Automated task reporting.** Like so many employees, project managers are often burdened with tedious tasks, from approving team members' work to sending check-ins if part of a project seems like it might run past a deadline. This is where AI comes in handy; by automating these tasks, a project manager can spend more time interacting with their colleagues.[49]
- **Resource allocation and prioritisation.** Project managers must be adept at deciding which parts of a project require the most equipment or labour, but these decisions are rarely straightforward. Luckily, AI-powered analytics can provide project managers with a wealth of information to make informed decisions in real time.

The project manager role will likely become one of the AI-affected roles that transforms rather than disappears. This is because, like most human-facing jobs, project managers still have to demonstrate nuance, empathy, and sensitivity when managing other employees, and as we've discussed, these skills will only become more valuable in an AI-powered workplace.

AI-Powered Tools for Project Management

Whether you are a project manager yourself or an employee looking to provide your project manager with a valuable workflow solution, you have several options for AI-powered project management tools:

- **Monday.com.** Project managers can use Monday.com

to automate repetitive tasks, gain insight into project progress, and auto-generate optimised workflows.

- **Trello.** Trello uses boards and cards to mimic a real-world idea board; the board is the space where all of a project's details are organised, and cards get added to the board to represent tasks and ideas. Trello's AI can automatically add cards to a project board based on defined project goals.
- **Asana.** Asana's AI integration can automatically remind an employee when they are near a deadline, and it can prioritise tasks in an employee's list based on preferences set by the project manager.
- **Wrike.** Wrike's Work Intelligence software can generate new content (like employee briefs) and adapt existing content to a given standard (ensuring company-wide consistency). It can also respond to voice commands, dramatically speeding up administration time.
- **Forecast:** An AI-powered research and project management tool, Forecast can proactively flag potential risks and suggest areas where a project manager could cut costs.

Whatever project management tool you or your company prefers, it probably comes equipped with AI-integrated tools to automate repetitive tasks and streamline workflow oversight.

Examples of AI in Communication, Team Management, and Project Coordination

Now that we've discussed some ways AI can contribute to communication and project management and the available tools, we can examine what it will look like for a company to use these tools in practice. Here, we will take six hypothetical workplace scenarios from a fictional company, the ABC Consulting Firm, and examine how AI integration can impact communication, collaboration, and project management.

AI-Powered Chatbots for Internal Communication

An employee for ABC Consulting is having trouble accessing a database, and they need IT support, but they are working late on a Friday night, and the company's IT staff are all off the clock. In a pre-AI world, the employee would have to send a message in the company Slack and hope that someone is available, or they would hold their question until the following Monday. However, in an AI-powered company, the employee could have their question addressed immediately by a chatbot. Chatbots can be set up to answer some of the most common employee queries, including HR policies and IT support issues.

Automated Meeting Scheduling

A manager at ABC Consulting is facing difficulty in organising an all-hands meeting due to the diverse schedules of their employees. They are unsure of the optimal time that would work for everyone. The manager can use an automatic meeting scheduling assistant to resolve this issue. This assistant is integrated into the company-wide calendar system and can analyse employee calendars to find a meeting time that fits most schedules. Suppose the assistant fails to find a time that suits everyone. In that case, the manager can set the AI to prioritise the schedules of two representatives from the marketing department who must attend the meeting.

Real-Time Language Translation

ABC Consulting has partnered with a firm in Japan, and they are going to work together on an upcoming project. However, ABC's team does not speak Japanese, which might cause some difficulty in communicating with their Japanese colleagues during the project planning process. To ensure that their colleagues feel comfortable and included, the team may decide to use translation tools during meetings or in written exchanges. AI-powered translation tools can instantly translate spoken or written text, making collaboration between the two companies efficient and smooth. AI-Enhanced Document Collaboration

An ABC Consulting project requires an extensive written report that all team members will complete. The team could use an AI-powered document collaboration tool to work together on the report in real time. The tool comes equipped with grammar edits and version control (if the team decides to revert to an earlier document version).

Predictive Analytics for Resource Allocation

A project manager for ABC Consulting has assembled a new team to work on a consulting project. It's a similar project to the ones they've worked on before, and because there are similarities, a bank of historical project data and information on team member skills are available. Thanks to this information, the project manager can automate the process of assigning team members to roles, putting more members on jobs that will take longer or putting more experienced members in more challenging roles. This automation speeds up the process of team members receiving their assignments, allowing them to start working sooner.

Task Management and Prioritisation

Now that the project manager has their project underway, they must ensure that the team members' various project parts stay on track. To do this, they use a workflow management tool like Asana or Trello, and they set it to automate specific tasks for their team members based on deadlines, dependencies, and team member availability.

Summary of Chapter Learnings

Communication and collaboration are essential components of any workplace. However, employees often feel overwhelmed with excessive and unnecessary communication. Fortunately, integrating AI into the workplace can help address these issues. For instance, AI can group individuals based on their job descriptions and interests and ensure they receive only relevant information. An internal chatbot can also be created to answer common employee queries, freeing up personnel to focus on more complex tasks. AI can also be used to generate presentations tailored to specific audiences, which can be beneficial for employees in communication-centric roles.

One of the advantages of AI-powered communication tools is that they are already integrated into many commonly used software, such as Google Workspace. Gmail, for example, includes AI-powered email composition and response writing suggestions. Google Drive has automated file sorting, while Zoom uses AI to auto-transcribe meetings, which can be shared with employees who were not present. Overall, integrating AI into the workplace can help improve communication and collaboration, increase productivity, and reduce workload.

Moreover, AI has the potential to support project management and communication. By leveraging AI algorithms, a project manager can build a team based on their availability, skills, and interests. Additionally, AI-powered project tracking can ensure that all team members are working on tasks with the highest priority and that the project is on track to meet the given deadline. In the realm of project management, software tools like Asana and Trello are quite popular, and they use AI to generate project deadlines and predict potential bottlenecks in a project.

At this point, we've covered many of the most common applications of AI in the workplace. Now it's time to zoom back out and once again take a higher-level look at AI as we discuss what it means to use AI ethically and responsibly.

10
THE ETHICAL AI EMPLOYEE: RESPONSIBLE USE OF AI

Imagine a scenario where a company relies solely on AI to make their crucial decisions, with no human oversight. In theory, the AI could make all the correct choices and drive the company towards exponential growth without requiring human input. However, this outcome is not guaranteed. As we've discussed, AI is not perfect. In a situation where it is given too much control over a company's decisions, it could also ruin the company if the bad decisions become reinforced in its algorithm. Furthermore, a company that relies too heavily on AI will likely find itself in a situation where job loss, rather than job transformation, is the reality.

The above scenario demonstrates the importance of ethical AI use. At the most fundamental level, AI ethics is the study, analysis, and discourse on reinforcing AI's beneficial impact while reducing risks. A solid grasp of ethical AI use is essential in an AI-powered workplace. Anyone using AI in their workplace needs to define exactly what the AI will be used for and how its results will be interpreted. This process needs to be conducted in such a way that AI complements the work that employees are doing rather than replacing employees entirely. Having an ethics-focused plan in place before you start utilising AI will be critical in ensuring that your AI integration maximises benefits and reduces risk.

Every AI ethics plan will look a little different depending on needs, interests, and goals. However, some basic principles would be an excellent starting point for any ethics plan:

- **Transparency.** Ethical use of AI requires both external and internal transparency. First, employees and companies should be open and honest about how they use AI. If you work in a client-facing position, inform your clients if you are collecting any data on them, explain how an AI algorithm will use their data, and describe the customer-side benefits. For internal communications, employees should make it known that reports, briefs, and emails were written either fully or partially by AI.[53]

- **Bias control.** When an employee uses AI to analyse data, it is essential to ensure that the data and the algorithm used are unbiased. Bias refers to any preconceived notions or opinions that can influence the results generated by the AI.[53] If the data or algorithm is biased, the AI will be trained to produce biased results, which can lead to inaccurate or unfair conclusions.[53] For example, suppose an employee is developing an eye-colour recognition software. In that case, they need to ensure that the data used to train the algorithm is not biased towards any particular eye colour. If the data contains a disproportionate number of brown eyes, the software will work better on brown eyes than on other eye colours. This bias can lead to inaccurate results when using the software on people with different eye colours.

- **Explainability.** Understanding how the algorithms work can be challenging if you are not actively involved in building or training AI algorithms. Too often, an employee or company will build an algorithm, feed the algorithm data, and report results. However, those who will benefit from the results need to understand how the AI produced those results. Anyone involved in building, training, or interpreting AI-generated products should either learn how to explain the entire AI-generation process in simple terms or train the AI to produce an explanation.[53]

- **Beneficence.** Before embarking on an AI integration plan, all parties involved should identify who will benefit from their machine learning entry and who may be affected negatively.[53] In assessing beneficence, it is crucial to consider beyond oneself - an employee may gain from increased productivity. Still, if such productivity comes at the expense of bias being integrated into machine learning, the employee should reconsider their approach to AI integration.

The Risks of Unethical AI Use

Sometimes, corners may be cut in the pursuit of remaining competitive. However, developing an ethical approach to AI integration is essential, as allowing unchecked AI operations can have severe consequences.

First, consider the risks of building unintentional bias into algorithms. A recent study conducted on the use of AI in recruitment efforts found that prospective employees who set their gender to "Female" were less likely to receive targeted advertising for higher-paying jobs than their male counterparts.[55] Using biased data to build recruiting algorithms can negatively impact diversity, equity, and inclusion efforts.

There are a few different types of bias to be aware of when building an algorithm:

- **Selection bias** occurs when the training data does not reflect the population it is supposed to serve.[57]
- **Algorithmic bias** in algorithms is reflective of the design choices made by their developers. If the developer unintentionally built bias into the algorithm, the algorithm will reflect that bias.
- **Confirmation bias** is a type of algorithmic bias. It describes a situation where the algorithm was designed (with good intentions) to look for something specific, but it reinforces stereotypes.[57] For example, a company using AI recruitment tools may encounter confirmation bias if the tools are designed to look for candidates who are similar to current employees.
- **Feedback loop bias** can occur even if the data and initial training are unbiased; if the algorithm receives biased feedback upon producing results, it can knock the algorithm out of calibration.[57]

Misinformation is another risk that needs to be mitigated. While some misinformation may not cause significant harm, such as ChatGPT getting a public figure's birthday wrong by a year, it can be dangerous regarding sensitive topics such as politics or religion. Therefore, it's crucial to take appropriate measures to mitigate such risks. Recently, OpenAI faced a lawsuit after ChatGPT fabricated allegations of embezzlement against a public figure. Of course, this does not mean that employees should suddenly stop trusting anything ChatGPT says; it is still, by and large, extremely accurate and efficient. But mistakes do happen, and it is especially important to double-check all facts, especially if they are of a sensitive nature.

AI users should look for potential copyright and property infringement issues. If content generation software returns material that is too similar to other content already under copyright or property protection, it can lead to a case of plagiarism or an infringement lawsuit if the user passes off the information as their own or misuses the copyrighted content.

Finally, AI users need to be prepared for cyberattacks and take steps to protect data and individual privacy. Given the ability of AI to generate code rapidly, some cybersecurity experts are concerned that bad actors could use AI to build bugs that can bypass security and access personal information. This means that employees who use AI should build protections into their AI ethics plan, which could include restricting who can access information and from which devices.

AI Ethics Examples and Solutions

How are companies currently handling AI ethics dilemmas? Here, we'll examine three real-world ethical problems that have arisen and discuss what companies are doing to address them.

Military Applications

As AI has evolved, companies have started discussing whether they will use AI for military applications if asked to do so.

At present, companies are coming down on both sides of the issue. For example, Google has taken a clear stance against AI use in military operations, saying that they will not use AI for "weapons or other technologies designed to cause or directly facilitate injury to people; in technologies that gather or use information for surveillance violating internationally accepted norms; or technologies for any purpose that contravene widely accepted principles of international law and human rights."[56] On the other hand, the software design company Palantir has already used AI to analyse stacks of data from defence and intelligence agencies; they hope to use their technology to make military operations more efficient and effective.[56]

Law Enforcement

AI tools like facial recognition have found their way into the law enforcement sector; agencies use them to identify potential suspects in open cases. However, some activists worry that the AI being deployed in law enforcement settings is biased, as it unfairly targets minorities. Brian Brackeen, the chief executive officer of facial recognition firm Kairos, argues that its usage "opens the door for gross misconduct by the morally corrupt."[56] Now, there are growing calls for AI companies to perfect their recognition software to minimise these biases and refrain from selling the software to law enforcement agencies until these agencies address their own systemic bias issues.[56]

Government Surveillance

In recent years, some nations have increased their surveillance activities, including those targeting their citizens. Actions taken by these nations include removing access to the Internet, imprisoning reporters, or targeting regime dissidents. As a result, some companies are now wary of selling their technologies to all government entities for fear of misuse. William Galston, who works for the American think tank the Brookings Institution, proposes a middle-ground solution in which government entities are given access to recognition software as long as the government operates under the presumption that it will not use the software unless given an ironclad justification for doing so.[56]

Even though these examples demonstrate the genuine and fundamental ethical issues surrounding the rise of AI, do not let these issues cause you to refrain from making AI a fundamental part of your workflow; AI still has tremendous potential to improve productivity and worker satisfaction. These examples are given only to provide context for the current state of AI ethics. Furthermore, with the right solutions in place, you can feel confident knowing that you are deploying AI ethically and responsibly.

As you develop an AI integration plan, here are five steps you (and your company) can take to mitigate associated risks:

- **Hire AI ethicists.** AI ethicists have made careers out of responsibly deploying AI, and having one or more on your team is essential to building your AI integration plan. They ensure that ethics are always part of the AI conversation and help companies instil an ethics-first culture.
- **Develop your AI ethics plan.** An AI ethicist can help you with this. All companies need to have a plan outlining who uses AI and how it is used, and all employees must completely abide by this plan. Furthermore, this plan should be readily available to the public for transparency.
- **Create an AI review board.** A team of employees representing all the sectors involved in the company should be assembled to regularly audit AI use and make recommendations.
- **Require AI users to leave a paper trail.** Anyone using AI in their workflow should be able to show in writing how the algorithm was designed and how the results are being interpreted. Having a written record of use is especially important if something goes awry with the AI and it results in a court case.

- **Implement ethics training into AI training programs.** Ethical use must be at the forefront of the program when teaching employees how to use AI.

The Future of Ethical AI in the Workplace

We are no longer asking "Can we?" regarding AI; now, the question has become, "How should we?" The ethical intersection of technology and humans requires plenty of forethought. Because AI is rapidly evolving, we need to consider how ethics and AI interact now and how they will interact in the future.

First, managers should start preparing to have conversations with their employees about how their roles will be redefined. Shifting job descriptions will become commonplace as AI continues to evolve, and managers need to start thinking about how they will adjust job descriptions and how they will communicate these changes to their employees. Furthermore, while very few jobs will be eliminated, that number will also not be zero; therefore, employers need to start figuring out what those jobs will be, preparing affected employees for the transition, and creating a system to support displaced workers.

In recent years, we have seen a shift in how we perceive the relationship between employees and employers and between employers and society. For example, more workforces are dispersed (participating in remote work), and more individuals are involved in self-employment and contract work rather than full-time employment. Recent studies have exposed the often difficult conditions in which these employees work. Therefore, companies with contract or self-employed workers need to start considering how they will use AI to ensure they are catering to their alternative workers, including using AI to monitor workload, stress levels, and potential threats to work-life balance. On the other side, self-employed workers should be prepared to speak up if they ever feel like AI is pushing them out.

The good news is that society is taking an increasing interest in ethics, and more companies are being held accountable for their ethical oversights. Customers and stakeholders will pay attention as companies integrate AI, and they will ask questions if they perceive that a company is not integrating AI responsibly or being transparent about the integration process. It bears repeating: every company should proactively prepare a detailed AI plan that explains exactly how the AI will be used, who will be using it, and what the benefits will be. If employees sense that their company is integrating AI without an ethics plan, they should take the initiative and introduce the idea to senior leadership. Additionally, they should insist that the plan be made readily available to the general public.

Summary of Chapter Learnings

Anyone looking to integrate AI into their workflow should first familiarise themselves with the fundamentals of ethical AI use. All AI integration should maximise benefits and minimise risks; this is the most fundamental principle of ethical AI use. As you or your company decide how to utilise AI, it is essential to design a plan that explains (in specific detail) what the AI will be used for, who will use it, and what the benefits will be.

The details of your AI ethics plan will be specific to your situation; however, a few basic tenets are valuable for almost any plan. These are transparency (ensuring that both employees and the public understand the situations in which you are using AI); bias control (ensuring that you are not building a biased algorithm or training an algorithm using biased data), explainability (ensuring that everyone involved in the AI's use understands how it is producing its results); and beneficence (analysing who will benefit and who could be harmed by the integration process).

Failing to prepare a well-defined ethics plan carries immense risk. Biased algorithms can reduce diversity, misinformation can lead to a libel suit, and copyright and plagiarism issues can arise if the AI user does not realise the algorithm has pulled copyrighted content. These risks reinforce the need for an ironclad ethics plan. One of the best first steps in developing a plan is to hire an ethics advisor; an advisor can ensure that any conversations about AI integration include ethical considerations, and they can foster a culture of ethical AI use within the company. Additional risk mitigation strategies include creating an AI review board, requiring AI users to leave a paper trail, and putting ethical considerations at the forefront of AI training programs.

Ethical considerations are not just a company-level issue; society is paying more attention to ethics than ever. This is another reason why companies need to develop a detailed, transparent ethics plan; clients and stakeholders will be watching, and they will speak up if they sense that ethically unsound practices are occurring.

Ethical AI use is one part of a broader, AI-ready mindset. In the next chapter, we'll look at another characteristic indicative of an AI-ready mindset: a willingness to embrace change.

11

EMBRACING CHANGE: CULTIVATING AN AI-READY MINDSET

"It is not the strongest species that survives, nor the most intelligent. It is the one that is the most adaptable to change."

—Charles Darwin

Adapting to Generative AI: Cultivating a Continuous Learning Mindset

Humanity's capacity for responding to technological advancements is nothing new; for all of human history, we have been a technology-forward species, constantly developing new tools to make our lives easier. In recent centuries, this process has sped up dramatically. Since the middle of the eighteenth century, we have gone through three major industrial revolutions: the mechanisation revolution, the communication and mobility revolution, and the digital revolution. Each of these revolutions required a massive shift in how we conceive of work. We are in the midst of a fourth revolution: the age of smart devices, high-speed networks, instant communications, and artificial intelligence. As with every technological revolution that has come before, this fourth revolution requires us to evaluate how these technologies will change the face of our workplace and adapt our workplace mentality accordingly.

As a professional, how can you stay relevant in the shifting AI landscape? The best thing you can do is develop your generative AI skill set. As you build your skills, you should be prepared to learn skills that compound your productivity in your existing job role (upskilling) or prepare you for a new job role (reskilling).

Upskilling and reskilling are poised to be essential parts of the future employee experience. According to the IBM Institute for Business Value, most executives anticipate a significant need for upskilling or reskilling, retraining up to 40% of their workforce within the next five years.[62] This rapid evolution means that many employees must be prepared to re-evaluate their job structures and skill requirements.

However, the nature of this skill shift will vary widely. While it's essential to develop basic literacy in all AI-centric skills, you can focus on upskilling and reskilling in certain skill subsets. For example, if you prefer computers, you may consider emphasising data literacy and coding skills. Proficiency in programming languages such as Python and R will help you work with AI algorithms and customise AI solutions to your specific needs without the need to hire an outside programmer.

Data literacy skills will be similarly essential to train the AI you have built on unbiased, high-quality data. Other employees may focus on critical thinking and problem-solving skills to interpret the results of an AI-generated analysis or identify opportunities where AI can further streamline processes and enhance productivity. Still, others may choose to emphasise building their human interaction skill set. This could include improvements in interfacing with clients, managing team members, or demonstrating AI's utility to stakeholders.

Then, there will be people who choose to focus on ethics skills, developing an awareness and an understanding of AI's implications in their workplace and advising the company on how to proceed with AI integration ethically and responsibly. As an employee, you will need to compare your current skill set, general strengths and weaknesses, and desired job description to determine how best to focus your upskilling or reskilling efforts.

Navigating Uncertainty: Building Resilience in a Generative AI-Driven Future

In the swiftly evolving landscape of generative AI, a certain amount of uncertainty is inevitable and understandable. However, if you cultivate an aptitude for resilience and flexibility, you will have a much easier time combating that uncertainty.

Shifting job descriptions is a prominent area of uncertainty for many employees. It's worthwhile to have a desired job description in mind as you develop your generative AI skill set, and it's essential to communicate with your employer as you prepare for a post-AI workplace. However, it's also vital to remember that your job description is the product of a relationship between you and your employer. You may find that work restructuring arises unexpectedly as your employer navigates the initial hurdles of AI integration. It's important to be as prepared for these unexpected changes as you are for the planned ones. Rather than being caught off guard by a surprising role change, prepare yourself mentally for the possibility that it could happen even if you have been in regular communication with your employer, and learn to embrace such changes with a growth mindset. Fostering a growth mindset enables you to embrace change and explore opportunities for professional development that you may not have had in your current role.

Even if you never find yourself surprised by a role change, you may face other AI-related uncertainties in your workplace, and as with shifting job descriptions, the best way to combat these uncertainties is by building resilience. Here are several strategies to help you cultivate a resilient mentality:

- **Embrace Lifelong Learning.** Adopt a mindset of continuous learning to stay informed of AI advancements and their implications for your role. Take advantage of professional development opportunities like workshops, online courses, or certifications (whether offered by your company or independently sought) that focus on AI technology. Investing in your learning journey will enhance your adaptability and enable you to leverage AI effectively in your work.

- **Stay Curious and Open-Minded.** Approach AI technology with curiosity and openness to new possibilities. Be receptive to exploring different AI applications and their potential impact on your workflow.

- **Experiment.** Take the initiative to experiment with new AI tools and techniques that could impact your productivity. Be willing to try new approaches and iterate based on feedback and outcomes. Remember that the learning curve may be steep initially, but if you keep at it, you will see results in time.

- **Foster collaboration and knowledge sharing.** Suppose you discover a practical new way to integrate AI into your workflow and share it with your colleagues. Conversely, pay attention if they share a new opportunity with you. By fostering a culture of knowledge sharing, you'll enhance adaptability for both yourself and your colleagues.

- **Develop Critical Thinking and Problem-Solving Skills.** These are transferable skills that will help you adapt to any role. Practice analysing complex problems and identifying creative solutions that leverage AI technology effectively. For example, if your company recently ran into an issue, think about how you would have solved the problem. Would you have made the same decisions the company made? Would your decisions have been impacted by using AI? Practising critical thinking and problem-solving skills can help you be a confident decision-maker when real scenarios arise.

- **Seek Feedback and Iterate.** These are valuable skills for any employee but are especially important when you start doing something new, like AI integration. Ask your colleagues whether your work has improved thanks to AI integration. Then, incorporate that feedback into your approach and re-evaluate your strategies based on it.
- **Cultivate Resilience and Agility.** Doing something new usually means making mistakes and experiencing setbacks. Know that these setbacks will happen, and embrace a growth mindset in which you view obstacles as opportunities for learning and growth. If you can cultivate the ability to pivot quickly in response to changing circumstances and emerging opportunities, you will be able to adapt more effectively to the evolving landscape of AI in the workplace.

Another thing you can do to assuage your uncertainness is to take an honest look at your current AI-related skill set; knowing your current AI readiness baseline will give you a concrete sense of what skills you need to address as you continue your AI journey.

Unveiling Your AI Readiness Baseline

One advantage of an AI readiness baseline is that it can help you create personalised learning pathways tailored to your specific needs and goals. By identifying areas for improvement, you can focus your efforts on acquiring skills that will have the most significant impact on your career advancement and job satisfaction. For example, if you work in a position that requires a lot of public speaking, you could take courses to improve your presentation skills and seek training on AI-powered presentation assistance tools.

With knowledge of your AI baseline, you can develop a strategic approach to skill enhancement that aligns with your career objectives. If you identify in-demand AI skills within your field, you can prioritise skill development to remain adaptable and competitive.

A clear understanding of your AI readiness also helps you make informed decisions regarding career paths, professional development opportunities, and integrating AI tools into your work. For example, you can use your AI baseline to assess whether a particular job opportunity aligns with your current skill set and future career aspirations. Additionally, you can evaluate the potential impact of integrating AI tools into your workflow and determine whether you need additional training to make the integration run smoothly.

Mapping AI to Your Professional Journey

No matter where you are in your AI journey, it is always a good time to pause and evaluate your AI readiness baseline—regular re-evaluation can ensure that your AI usage is mapped to your professional goals. If you've been working with AI for a while, you may already have an established process to gauge your baseline. However, if you are an AI newcomer, you will want to create a structured approach by which to assess your baseline. The AI Integration Starter Assessment for Newcomers (found in the back of the book) is a wonderful resource. It was created to guide professionals to incorporate AI into their work, starting with foundational knowledge and recognising AI in everyday contexts.

In this assessment, you start with a simple exercise: defining what AI means to you and exploring how you currently use AI in your daily life. You will then move on to listing your daily tasks and assessing how AI tools can assist with each of those tasks (a list of example tools is available on assessment; feel free to draw from this list or bring in your own tools). Next, you will assess your AI knowledge on a 0-10 scale and set detailed learning goals. Finally, you will craft an AI learning path based on your goals and identify a task with which to begin your AI integration journey, choosing a task that would be immediately beneficial and outlining the steps it will take to integrate AI into the task. If you are new to AI, a structured approach like this will be invaluable to creating tangible results from AI integration.

Summary of Chapter Learnings

Thanks to new technological advances, humans are characterised by their adaptability, specifically their ability to learn and grow. In the past few centuries, humans have gone through three major technological revolutions that have seen the development of mechanisation, communication, and digitisation. Now, we have entered a fourth revolution characterised by smart devices, high-speed networks, instant communications, and artificial intelligence. In such a fast-evolving technological landscape, embracing lifelong learning and adaptability becomes essential for staying relevant in the workforce.

One of the most essential ways for professionals to stay relevant in the shifting AI landscape is to develop their generative AI skill set through upskilling and reskilling. Upskilling is defined as learning new skills that enhance your productivity in your existing job role, while reskilling is building a skillset in preparation for a new role. An AI-savvy employee should embrace adaptability by both upskilling and reskilling, especially because executives anticipate a significant need for upskilling or reskilling a substantial portion of the workforce due to generative AI within the next five years.

Fundamental literacy in various skills will make you a desirable employee; however, you should also consider your strengths, areas for improvement, and desired job description and focus on AI-complementary skills in relevant areas. For example, if you prefer working directly with clients, you might develop basic data literacy and then focus extensively on honing your human interfacing skills. Other focus areas include data literacy, coding, critical thinking, problem-solving, and ethics.

Even though AI has become increasingly ubiquitous, uncertainty about its use still looms. However, an employee who learns adaptability and resilience will have an easier time assuaging these concerns. Anyone interested in AI integration should embrace lifelong learning, stay curious and open-minded, develop an experimental mindset, emphasise collaboration, develop critical thinking and problem-solving skills, and seek feedback.

Another factor in combating uncertainty is self-reflection. Understanding your comfort level with AI will give you a concrete sense of how far you need to go to become an AI-savvy employee. An AI readiness baseline is an excellent tool for such self-reflection. An AI-readiness tool will help you create personalised learning pathways, formulate strategic skill enhancement approaches, make informed decisions, and engage in proactive career planning. Understanding your AI readiness baseline allows for a structured approach to assessing and improving AI-related skills, ensuring alignment with your career objectives and professional goals.

You will find the AI Integration Starter Assessment for Newcomers at the back of the book. This is a resource for professionals embarking on their AI integration journey, and it will give you a structured approach to gauge your AI readiness. The assessment starts with the basics, asking you to decide what AI means to you and list how you already use AI in your life. You will then list daily tasks AI could automate, evaluate your current AI knowledge, set learning goals, craft an AI learning path, and identify a task for immediate AI integration. The AI Integration Starter Assessment is a great first step as you work toward becoming an adaptable, resilient employee with an AI-ready mindset.

Now, as we're reaching the end of the book, it's time to look into the future. What will future AI-driven workplaces look like, and how can we prepare for them?

12

AI-DRIVEN FUTURES: NAVIGATING NEW CAREERS AND COLLABORATIONS

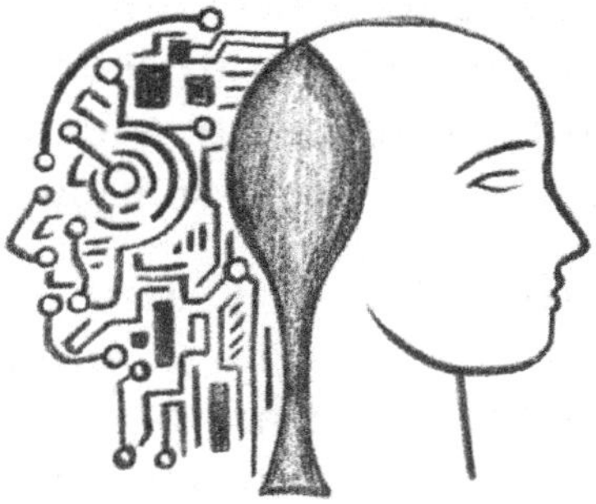

The future of generative AI is bright. As researchers continue to break new ground in developing AI modules, analysts are looking into the future, attempting to determine the most important AI trends of the next 5-10 years. Of course, looking into the future requires a certain amount of guesswork, and it's impossible to know exactly what AI development will emerge as the next big thing; it may be that the next worldwide phenomenon, like ChatGPT, is something that we cannot even conceive of yet. However, analysts have thoughts about the most exciting AI frontiers and how they will be conquered in the next few years.

At present, most AI only works with one input type. An algorithm can take in text, audio, or visual information, but not all three at the same time; some algorithms have been developed to convert text to images or speech into text, but no algorithm can yet move freely between data types and complete a variety of tasks. However, research into multimodal AI may allow an algorithm to source information from multiple data types and move freely between various data types based on the user's needs. For example, a user might be able to ask a question about a video and receive a natural language answer, or they might ask out loud for a recipe and receive both a written step-by-step guide and visual instructions. Multimodal AI will also lead to advances in video as a data input, which could give a company more power to track client engagement or further automate the hiring process. Some introductory multimodal models have been introduced, like OpenAI's GPT-4V and Google's Gemini, which can move between natural language processing and computer vision tasks.[65] Google also recently announced a new program called Lumiere which will have image-to-video capability and will be able to take in images as data for style references.[65]

Sam Altman, the CEO of OpenAI, has suggested that the future direction for language models is one of size reduction.[65] The recent trend has been for language models to get bigger and bigger, requiring massive amounts of information and correlated massive access to energy; in one day, ChatGPT uses the same amount of energy as 33,000 households.[65] Altman believes the models have gotten so big that we have reached the point of diminishing returns and that future language models will be improved in other, less resource-intensive ways.[65] Therefore, researchers in language modelling are focusing on deriving high-quality results with less resource intensiveness. Less resource-intensive AI modules will empower small businesses to run, train, and develop the algorithms they need for their business without needing access to a cloud-based service or a huge source of computing power.[65] Additionally, running smaller models on local devices mitigates data privacy issues; keeping sensitive data out of the cloud makes it harder for hackers to access it.

In general, the future trend of AI is all about accessibility, and one of the easiest ways to make something accessible is to make it open-source. New AI development techniques and resources will be available for free to the average consumer. Open-source datasets are becoming more widespread; these datasets are available to anyone to create a base layer of training or fine-tune an AI model.

Throughout this book, we've talked extensively about

the utility of a virtual agent to automate certain tasks in your life; these virtual agents will only become more powerful with advances in technology. In particular, virtual agents will integrate with multimodal AI. For example, an employee may be able to take a video from a project site and ask the AI to analyse resource needs based on the video, or a business executive may ask their virtual assistant to research and book the best flight from London to New York for a conference (at present AI can research the best options for you, but it cannot directly book your travel).

Evolving Work Patterns and AI's Role

With so many exciting developments in the works, AI will transform how we live over the next few years, and that change will be felt especially in the workplace. In one way, AI is going to reinforce trends that we already see in our workplaces; in another way, AI will create entirely new trends.

Today, having a full-time job no longer requires commuting to a central office. An increasing number of employees are logging in remotely, whether from their own home or a coworking space. The COVID-19 pandemic may have prompted this change, but it has staying power thanks to AI. As discussed in Chapter 9, AI is making workplace communication easier than ever, especially when dealing with a remote workforce. With AI integration, a remote company can automatically schedule Zoom meetings for employees logging in from across time zones or integrate workplace tools that allow for real-time collaboration.

Like remote companies, the rise of dispersed companies (those with multiple offices) has been possible thanks to the AI revolution. Whether a company is dispersed or fully remote, AI communication tools enable it to function with its employees, who are spread across the globe. As a future-focused employment trend, remote and dispersed workforces promote increased employee satisfaction and expand a company's reach, which is essential in an era of increased globalisation.[69]

Knowledge work is another workplace paradigm that will change thanks to AI. We've discussed some ways that AI will shape the future of knowledge work, which relies on critical thinking, decision-making, and collaboration skills. However, recent research shows that significant knowledge work automation will likely occur sooner than previously thought. This is because generative AI's natural language processing capabilities have evolved faster than expected.[64] Therefore, people in knowledge-based fields like education, law, technology, and the arts should start developing their AI-complementary skills now because the time when AI automates significant parts of their work is likely only a couple of years away.

Typically, the focus on AI in the workplace is related to productivity gains; in other words, how AI can help current employees become more productive and, thereby, more satisfied in their roles. However, one often-overlooked benefit of the AI revolution is that it will create entirely new jobs. These new jobs will offset potential losses caused by job automation and will give new entrants to the job market an opportunity to help build a future-focused workplace.

The most obvious jobs AI will create are those directly related to the AI itself. Algorithm builders will design the algorithms on which AI is founded, trainers will fine-tune the algorithms to meet the expected outcomes, IT staff will maintain the hardware and software infrastructure, teachers will train new employees how to work with AI, and ethicists will review AI use in the workplace and make recommendations.

However, opportunities not directly related to AI building, training, and maintenance will also arise. Many of these jobs will be defined by how the employee utilises AI. For example, sentiment analysers will use generative AI to sift through blogs, reviews, and social media posts to determine how consumers feel about a company's products and services.[66] AI can compile feedback, but human intuition still needs to interpret that feedback using emotional and contextual knowledge. Creatives will also find new ways to interact with AI. One area in which AI has grown exponentially is virtual reality, with tools like gesture recognition, emotion recognition, and AI-driven character animation and behaviour systems.[66] However, virtual reality AI will still require humans to lay the groundwork for the worlds that AI renders; humans will design the virtual worlds, characters, simulations, and experiences that AI will then render and enhance.[66] Writers and influencers will indirectly help produce better-trained AI; as content creators and influences produce engaging content, AI will absorb that content and produce similarly compelling work, improving the occasionally bland responses that AI currently produces.[66] AI has also become capable of producing artistic works like prose, pictures, and video; however, it still needs to be given a vision with which to create the content. AI art directors will create the vision for a project and then work with the AI to produce the work, adjusting the AI's directives as necessary.[66]

Augmented Intelligence: The Future of Human-AI Collaboration

The future of AI is one of collaboration; humans and AI will work in tandem to improve the workplace experience. As we prepare for this collaborative future, it's important to have a vision of a successful AI-integrated workplace. A successfully integrated workplace will have clearly defined goals for the AI-human collaboration, an understanding of the degree of agency the AI will have, and a plan to deal with the consequences if the AI fails to perform as expected. Of course, this only scratches the surface of the considerations that must be made when integrating AI.

The best way to visualise a successful AI-integrated workplace is to look at three evolving workplace scenarios, specifically how they integrate AI and address some of the questions that are part of that process.

Educators are employing intelligent tutoring systems (ITSs) to give students additional one-on-one attention.[68] ITSs are designed to be both educational and entertaining as they guide students in developing different types of knowledge—factual, procedural, problem-solving, etc.[68] Advanced ITSs can also read students' speech and facial expressions and adjust their approach based on perceived student engagement.[68] For educators, ITSs are an extension of the learning experience; they can help ensure that students stay engaged outside the classroom when a teacher is absent.[68] However, while the ITS is responsible for direct student engagement, the teacher will train the ITS to ensure that it tutors the students on the correct material, continuously monitors student progress, and intervenes as necessary.

Artists know that good drawings take a long time, especially with an initial sketch. This is where AI-powered drawing tools show their might; AI systems like Sketch-RNN and Magic Sketchpad can advance or complete sketch drawings that the user starts, and they can be trained to mimic one's own drawing style.[68] In some cases, the AI may take the sketch in an unexpected direction that the user appreciates, in essence making the AI a collaborator in the idea generation process. In a successful artist-AI partnership, the artist will develop the initial vision and provide the base drawing style; then, as the AI finishes the sketch, the artist will provide additional input or override the AI's decisions if they disagree with the direction the AI has taken the sketch.[68]

Doctors use magnetic resonance imaging (MRI) to detect tumours and lesions in body tissues and organs.[68] The quality of MRI images has been inconsistent for a long time, leading doctors to interpret images with less confidence.[68] However, convolutional neural networks specialising in image processing can produce clearer images more efficiently than traditional methods.[68] Recent developments have allowed AI to start processing images themselves, flagging potential areas of concern for human review.[68] In this case, doctors still need to check the AI's work because it is especially important to maintain a human touch in an emotionally sensitive field with significant consequences of failure like medical imaging. The emphasis in all three scenarios is on collaboration; however, it is a collaborative atmosphere where the human provides the impetus for the project and has the final say on whether to accept or reject the AI's decisions.

Preparing for an AI-Driven Future

AI is here to stay; its evolution will inevitably shape our future. Preparing for an AI-driven future will require two significant steps: concrete strategic planning and a mindset shift. Concrete strategic planning refers to the tangible actions you can take to improve your AI-savviness. Conduct an AI readiness baseline assessment, seek out AI-centric continuing education opportunities, begin experimenting with AI tools to improve your workflow, develop AI-adjacent skills like empathy, critical thinking, and problem-solving, and familiarise yourself with the principles of AI ethics.

The first and most important part of the mindset shift is to embrace an AI-centric reality. Know that AI is the now and the future, and choose to accept it into your lifestyle rather than fighting against it. By just choosing to embrace AI, you will position yourself to enter the AI revolution at a time when you can still be proactive in its uptake rather than reactive.

Adjacent to the choice to embrace AI is the choice to embrace resiliency, adaptability, and a growth-first mindset. Integrating AI will almost certainly entail a learning curve, but if you view the learning curve as an opportunity to better yourself, you will eventually reap tremendous benefits. As you learn, understand that you will face hurdles and probably make mistakes. This is where a resilient mindset comes into play; if you focus on getting past and learning from those hurdles, you will eventually reach the point where you are beyond the most challenging part of the learning curve. Finally, become a constant student of AI, stay abreast of the latest developments, and be prepared to make changes when new technology emerges that impacts your current workflow.

Summary of Chapter Learnings

The future of generative AI is bright; researchers and analysts are already speculating about the amazing developments we may see in the next five to ten years. One promising area of research is multimodal AI. Currently, AI typically operates within single input types, such as text, audio, or visual data. However, multimodal AI can operate freely across input types, allowing for greater flexibility in how humans and AI collaborate. Analysts are also looking toward smaller, more resource-efficient language models, open-source development, and increasing availability to the general public as promising trends in AI research.

AI's integration into the workplace both reflects and informs current trends in workplace dynamics. For example, remote and dispersed workforces became popular during the COVID-19 pandemic, but they have had staying power thanks to AI-driven communication tools. Tools like workflow management and real-time collaboration software allow employees to maintain productivity while choosing a work environment that suits their personal needs. Additionally, the automation of knowledge work, driven by advancements in natural language processing, will require employees in knowledge-based fields to develop AI-complementary skills. Recent research shows a faster-than-expected uptake of AI, so these employees may need to build their AI skillset sooner than anticipated.

The AI revolution will create new job opportunities. These new roles will be directly related to AI development, training, maintenance, ethics, etc. But beyond these roles, AI will create new jobs informed by how employees use AI. Sentiment analysts will work with generative AI to parse through client feedback, and creatives will collaborate with AI in fields like virtual reality and art production to make their visions come to life.

The future of AI is one of collaboration; humans and AI will work together to improve workplace productivity. In a successful collaboration, humans will set clear goals and defined roles for the AI, monitor the AI's results, and reserve the right to make decisions or retrain the AI to fit their needs.

Preparing for an AI-driven future involves both strategic planning and a mindset shift toward embracing AI and fostering resilience and adaptability. Embracing AI will require lifelong learning and a willingness to adapt to emerging technologies. By preparing both physically and mentally, you will be ready to thrive in an AI-powered future.

CONCLUSION

As our journey nears its end, let's take some time to look back. We started with the basics, discussing the current state of AI and its growing impact on the modern workplace and overviewing the history and basic functionality of generative AI. We also learned how AI can boost productivity, including automation of routine tasks, time management, content creation and workflow optimisation, and we addressed some of the common myths and misconceptions surrounding AI. After that, we transitioned from theoretical to practical applications, learning to harness the power of ChatGPT and other tools to help with automation, optimisation, decision-making, and collaboration. We also took a brief dive into financial assistance, looking at how AI can improve your financial stability and assist you in building a solid stock portfolio. Having gone through these practical chapters, we returned to theoretical applications as we discussed ethical AI use and the importance of embracing change and cultivating an AI-ready mindset. Finally, we looked into the future and discussed some emerging technologies that will define AI within the next decade.

We covered a lot of ground in this book, and you're probably now brimming with ideas and knowledge. But let's take a second and review the core points, the essential takeaway messages you will want to remember long after you reach the end of this book.

- In the broadest sense, artificial intelligence is defined as human-like intelligence represented by machines. Artificially intelligent machines can perform specific actions similar to and, in some cases, better than humans.
- The most significant benefit of artificial intelligence in the workplace is its ability to boost your productivity. Productivity boosts can come in the form of routine task automation, time management, content creation, decision-making assistance, and workflow optimisation—the possibilities are limited only by your imagination.

- While a few jobs will likely be lost due to AI automation, most jobs will simply transform. Additionally, new jobs will be created that support and reflect AI's evolution.
- AI is already being used in most sectors. Companies are using AI to revolutionise the shopping experience, conduct project management and forecasting, build predictive maintenance schedules, and so on. AI has the power to revolutionise almost any industry.
- The AI flywheel metaphor is a great way to conceptualise AI integration. When you incorporate AI into your operations, you'll set in motion a perpetual cycle in which integration leads to greater productivity, productivity leads to new opportunities, opportunities lead to increased confidence, and confidence leads to further AI integration.
- It's possible to personalise ChatGPT to meet your requirements and teach yourself how to use ChatGPT commands that will yield better outcomes. Additionally, remember that ChatGPT is not the only AI resource available. There is a plethora of software options, ranging from free to professional-grade models, that can assist with various aspects of your workflow.
- It's essential to use AI ethically and responsibly. As you integrate AI into your workflow, develop an ethics plan so you know exactly how AI will be used, how its efficacy will be evaluated, and who will benefit from its use.
- AI readiness is a mindset. To embrace AI, you must become a lifelong learner, always ready to adopt new technologies as they emerge. You will also need to constantly evaluate your skillset, upskill or reskill as necessary, cultivate resilience, and adapt to the fast pace of change in an AI-driven workplace.

AI is the future of the professional world. Exciting new developments in AI are on the horizon. Multimodal AI will allow users to move freely between data types in their AI interactions, and advances in natural language processing will make virtual assistants like ChatGPT less resource-intensive and more accessible to average consumers, as will the rise of open-source AI technology. The impact of AI will continue to be felt across industries, and it will continue to inform the nature of the workplace, especially as it promotes remote and dispersed workforces. In general, the future of AI in the workplace is one of close collaboration with humans. An AI-driven future is near, so employees should start preparing by developing AI-complementary skills like computer literacy, data analysis, decision-making, critical thinking, and empathy. Employees should also settle into their AI-ready mindset, embracing resilience, adaptability, and lifelong learning.

This book begins your AI journey but should not be the end. As someone who has embraced lifelong learning, you should seek additional resources to continue your AI education. Companies like Coursera, edX, and Udacity offer courses like "Machine Learning," "Deep Learning Specialisation," or "AI Programming with Python" through which you could gain a deeper understanding of how AI works. Forums and communities like Stack Overflow, Data Science Central, and Towards Data Science offer articles, forum posts, and discussion spaces where you can learn about the latest AI advancements at your own pace or post questions. Professional networks like the Association for the Advancement of Artificial Intelligence and LinkedIn are great places to network with like-minded AI users and engage in a supportive community.

The AI revolution presents an incredible opportunity for you as an employee. Integrating AI into your workflow can increase your workplace productivity, opening doors to new possibilities and avenues for growth. If you have ever felt overwhelmed by your workload, afraid that advancing technology will leave you behind, frustrated by stagnation in your career or income, wanted more free time to spend with your loved ones, or wanted to be part of a revolution, it's time to embrace AI.

Albert Einstein once said, *"The important thing is not to*

stop questioning. Curiosity has its own reason for existing." This is what it means to embrace AI. If you continuously seek knowledge, question the status quo, and explore the untapped potential of AI in transforming your career and enhancing your productivity, you will future-proof your career and prepare yourself to succeed in an AI-driven world.

NOTES

Chapter 1.

1. Amazon Staff. (2023, December 7). A New Study Reveals 5 Ways AI Will Transform the Workplace as We Know It. About Amazon.https://www.aboutamazon.com/news/aws/how-ai-changes-workplaces-aws-report

2. Bannon, M. T. (2023, June 22). How AI Is Changing The Future Of Work. Forbes. https://www.forbes.com/sites/marenbannon/2023/06/22/how-ai-is-changing-the-future-of-work/?sh=43dab68a72e0

3. Marr, B. (2023, June 27). Boost Your Productivity with Generative AI. Harvard Business Review. https://hbr.org/2023/06/boost-your-productivity-with-generative-ai

4. Noenickx, C. (2023, July 13). Workplace AI: How Artificial Intelligence Will Transform the Workday. BBC. https://www.bbc.com/worklife/article/20230515-workplace-ai-how-artificial-intelligence-will-transform-the-workday

5. Semuels, A. (2020, August 6). Machines and jobs: How the coronavirus has accelerated automation. Time. https://time.com/5876604/machines-jobs-coronavirus/

6. Takyar, A. (n.d.). AI Use Cases & Applications Across Major Industries. LeewayHertz. https://www.leewayhertz.com/ai-use-cases-and-applications/#:~:text=AI%20in%20business%20intelligence%20automates,code%20quality%20through%20automated%20reviews.

Chapter 2.

7. Coursera Staff. (2023, November 29). 4 Types of AI: Getting to Know Artificial Intelligence. Coursera. https://www.coursera.org/articles/types-of-ai

8. Hardesty, L. (2017, April 14). Explained: Neural Networks. MIT News.https://news.mit.edu/2017/explained-neural-networks-deep-learning-0414

9. IBM Data and AI Team. (2023, October 12). Understanding the different types of artificial intelligence. IBM Blog. https://www.ibm.com/blog/understanding-the-different-types-of-artificial-intelligence/

10. Roose, K. (2022, December 5). The Brilliance and Weirdness of ChatGPT. The New York Times. https://www.nytimes.com/2022/12/05/technology/chatgpt-ai-twitter.html

Chapter 3.

11. Carothers, D. (2023, March 20). Leveraging AI and ChatGPT for the Modern Insurance Agent. HawkTalk Blog. https://blog.hawksoft.com/chatgpt-applications-insurance#risk

12. Chinn, A. (2023, May 17). The HubSpot Blog's State of AI Report [Key Findings from 1000+ Marketing Pros]. The HubSpot Blog. https://blog.hubspot.com/marketing/state-of-ai-report?hubs_content=www.hubspot.com%2Fproducts%2Fartificial-intelligence&hubs_content-cta=The%20HubSpot%20Blog%27s%20State%20of%20AI%20Report&_ga=2.127740557.1042228796.1706691005-1547609410.1706691005

13. Noy, S. & Whitney Zhang. (2023, March 2). Experimental Evidence on the Productivity Effects of Generative Artificial Intelligence. (Working Paper). https://economics.mit.edu/sites/default/files/inline-files/Noy_Zhang_1.pdf

14. Owens, R. (2023, November 30). The 10 Best AI Data Analysis Tools in 2024. Polymer. https://www.polymersearch.com/blog/the-best-10-ai-tools-to-analyze-data

15. Kudalkar, D. (2024, March 20). TacticAI: Google DeepMind's answer to next-gen football strategy. FavTutor. https://favtutor.com/articles/tacticai-google-football-assistant/

Chapter 4.

16. Bughin, J., Hazan, E., Lund, S., Dahlstrom, P., Weisinger, A., & Subramaniam, A. (2018, May 23). Skill shift: Automation and the future of the workforce. McKinsey Global Institute. https://www.mckinsey.com/featured-insights/future-of-work/skill-shift-automation-and-the-future-of-the-workforce

17. Callahan, C. (2023, June 5). Here are Artificial Intelligence's Biggest Misconceptions. WorkLife. https://www.worklife.news/technology/ai-mythbuster/

18. Chakravorti, B. (2023, June 25). How will AI change work? A look back at the 'productivity paradox' of the computer age shows it won't be so simple. The Conversation. https://thecon-

versation.com/how-will-ai-affect-workers-tech-waves-of-the-past-show-how-unpredictable-the-path-can-be-205734

19. Editors Desk. (2024, February 28). The AI Paradox: Balancing Automation and Employment in the Future of Work. Analytics Week.https://analyticsweek.com/the-ai-paradox-balancing-automation-and-employment-in-the-future-of-work/#:~:text=The%20'AI%20paradox'%20lies%20in-,swathes%20of%20the%20workforce%20adrift.

20. Gibbons, S. (2023, December 7). How AI Might Impact The Job Market In 2024. Forbes. https://www.forbes.com/sites/serenitygibbons/2023/12/07/how-ai-might-impact-the-job-market-in-2024/?sh=5991873a461d

21. Kande, M., & Likens, S. (n.d.). 6 generative AI business myths that will make you rethink everything. PwC. https://www.pwc.com/us/en/tech-effect/ai-analytics/six-generative-ai-business-myths.html

22. PwC. (2022, May 25). Hopes and fears 2022: Global workforce survey. PwC. https://www.pwc.com/gx/en/issues/workforce/hopes-and-fears-2022.html

Chapter 5.

23. Podean (n.d.). Amazon Explained – The Flywheel. Podean. https://www.podean.com/amazon-explained-the-flywheel/

24. Dell'Acqua, F., III, E. M., Mollick, E., Lifshitz-Assaf, H., Kellogg, K. C., Rajendran, S., Krayer, L., Candelon, F., & Lakhani, K. R. (2023, September 22). Navigating the Jagged Technological Frontier: Field Experimental Evidence of the Effects of AI on Knowledge Worker Productivity and Quality. Harvard Business School Working Paper No. 24–013.

25. Nielsen, J. (2023, July 16). AI Improves Employee Productivity by 66%. Nielsen Norman Group. https://www.nngroup.com/articles/ai-tools-productivity-gains/

26. Underhill, J. (2024, February 29). Case studies and success stories: AI in project management. LinkedIn. https://www.linkedin.com/pulse/case-studies-success-stories-ai-project-management-underhill-ppm--axtmf/

27. Peerbocus, S. (2024, May 13). Embracing the future: How AI agents are revolutionizing project management. LinkedIn. https://www.linkedin.com/pulse/embrac-

ing-future-ai-agents-revolutionizing-project-peerbo-cus-pmp-yud9f/

28. Gandhi, S. (2024, June 10). Case study: How Uber uses AI to optimize surge pricing. LinkedIn. https://www.linkedin.com/pulse/case-study-how-uber-uses-ai-optimize-surge-pricing-shripal-gandhi-s54mf/

29. Bandoim, L. (2019, April 30). How McDonald's plans to use machine learning to personalize the drive-thru. Forbes. https://www.forbes.com/sites/lanabandoim/2019/04/30/how-mcdonalds-plans-to-use-machine-learning-to-personalize-the-drive-thru/

Chapter 6.

30. Kreacic, A., Lasater-Wille, A., Uribe, L., Jesuthastan, R., Romeo, J., & Luong, S. (2024). How Generative AI Is Changing The Future Of Work. https://www.oliverwymanforum.com/global-consumer-sentiment/how-will-ai-affect-global-economics/workforce.html

31. Lau, J. (2024, February 21). How to use ChatGPT. Zapier. https://zapier.com/blog/how-to-use-chatgpt/

32. Ray, P. P. (2023, April 14). ChatGPT: A comprehensive review on background, applications, key challenges, bias, ethics, limitations and future scope. Internet of Things and Cyber-Physical Systems, 3, 121–154. https://doi.org/10.1016/j.iotcps.2023.04.003

33. Jarow, O. (2023, March 5). How the first chatbot predicted the dangers of AI more than 50 years ago. Vox. https://www.vox.com/future-perfect/23617185/ai-chatbots-eliza-chatgpt-bing-sydney-artificial-intelligence-history

34. OpenAI. (n.d.). ChatGPT pricing. OpenAI. https://openai.com/chatgpt/pricing/

Chapter 7.

35. Dennison, K. (2023, March 14). The Impact Of Artificial Intelligence On Leadership: How To Leverage AI To Improve Decision-Making. Forbes. https://www.forbes.com/sites/karadennison/2023/03/14/the-impact-of-artificial-intelligence-on-leadership-how-to-leverage-ai-to-improve-decision-making/?sh=49ce21e533d9

36. Oracle. (2023, April 19). Global Study: 70% of Business Leaders Would Prefer a Robot to Make Their Decisions. PR Newswire. https://www.prnewswire.com/news-releases/global-study-70-of-business-leaders-would-prefer-a-robot-to-make-their-decisions-301799591.html

37. Petrova, I. (n.d.). AI-enabled Decision-making: Enhancing Productivity and Agility in Organizations. EVALUESERVE. https://www.evalueserve.com/blog/ai-enabled-decision-making-enhancing-productivity-and-agility-in-oganizations/#:~:-text=Human%2Din%2Dthe%2Dloop,outcomes%20and%20 intervene%20when%20necessary.

38. Purdy, M., & Williams, A. M. (2023, October 26). How AI Can Help Leaders Make Better Decisions Under Pressure. Harvard Business Review. https://hbr.org/2023/10/how-ai-can-help-leaders-make-better-decisions-under-pressure

39. Wilson, J. H., & Daugherty, P. R. (2018, August). Collaborative Intelligence: Humans and AI Are Joining Forces. Harvard Business Review. https://hbr.org/2018/07/collaborative-intelligence-humans-and-ai-are-joining-forces

Chapter 8.

40. Abbamonte, K. (2023, April 13). AI job search tips: 9 AI tools to help you land your next job. Zapier. https://zapier.com/blog/ai-job-search/

41. Amazon Staff. (2023, December 7). A new study reveals 5 ways AI will transform the workplace as we know it. About Amazon. https://www.aboutamazon.com/news/aws/how-ai-changes-workplaces-aws-report

42. Benson, A. (2024, March 22). 6 Best Free Stock Screeners. Nerdwallet.https://www.nerdwallet.com/article/investing/best-stock-screeners

43. Carter, T. (2023, October 29). These skills could earn you more money and help futureproof your career. Business Insider. https://www.businessinsider.com/ai-skills-boost-salary-futureproof-career-study-2023-10

44. Grossman, A. (2024, January 7). How to Use Artificial Intelligence in Your Portfolio in 2024. Investopedia. https://www.investopedia.com/how-to-use-artificial-intelligence-in-your-investing-7973810

45. Monash University. (2023, December). How to equip yourself for AI career success. Monash Online. https://online.monash.edu/news/how-to-equip-yourself-for-ai-career-success/

46. Kinderlin, S. (2024, January 10). Is AI impacting your job? Your salary might actually rise, top talent CEO says. CNBC. https://www.cnbc.com/2024/01/10/ai-impacting-jobs-could-lead-to-higher-salaries-top-talent-ceo-says.html#:~:text=Work-,Is%20AI%20impacting%20your%20job%3F,rise%2C%20top%20talent%20CEO%20says&text=The%20artificial%20intelligence%20boom%20has,lower%20skill%20level%20from%20employees

47. University of San Diego (n.d.). The Impact of AI on the Job Market and Employment Opportunities. University of San Diego Online. https://onlinedegrees.sandiego.edu/ai-impact-on-job-market/#:~:text=AI%20affects%20jobs%20by%20altering,potential%20biases%20and%20privacy%20concerns.

Chapter 9.

48. Grensing-Pophal, L. (2023, October 17). Using AI to Enhance Employee Communications. SHRM. https://www.shrm.org/topics-tools/news/technology/using-ai-enhance-employee-communications

49. Champlain College Online. (2024, January 19). How Artificial Intelligence Is Revolutionizing Project Management. Champlain College Online. Online.https://online.champlain.edu/blog/how-artificial-intelligence-revolutionizing-project-management#:~:text=AI%27s%20Role%20in%20Enhancing%20Efficiency&text=Firstly%2C%20it%20will%20automate%20minimally,complex%20insights%20through%20machine%20learning.

50. Lau, G. (n.d.). AI in communication: How businesses are using AI to improve internal and customer communications. Dialpad. https://www.dialpad.com/blog/ai-in-communications/

51. PMI. (2023, October). Shaping the Future of Project Management With AI. PMI. https://www.pmi.org/learning/thought-leadership/ai-impact/shaping-the-future-of-project-management-with-ai

52. Corporate English Solutions. (2023, October 16). Transforming workplace language and communication: Introducing AI. Corporate English Solutions. https://corporate.britishcouncil.org/insights/transforming-workplace-language-and-com-

munication-introducing-ai

Chapter 10.

53. Marr, B. (2021, September 10). How Do We Use Artificial Intelligence Ethically? Forbes. https://www.forbes.com/sites/bernardmarr/2021/09/10/how-do-we-use-artificial-intelligence-ethically/?sh=42192f8279fd

54. Schwartz, J., Denny, B., & Mallon, D. (2020, May 15). Ethics and the future of work. Deloitte Insights. https://www2.deloitte.com/us/en/insights/focus/human-capital-trends/2020/ethical-implications-of-ai.html

55. Spisak, B., Rosenberg, L. B., & Beilby, M. (2023, June 30). 13 Principles for Using AI Responsibly. Harvard Business Review. https://hbr.org/2023/06/13-principles-for-using-ai-responsibly

56. West, D. (2018, September 13). The role of corporations in addressing AI's ethical dilemmas. Brookings. https://www.brookings.edu/articles/how-to-address-ai-ethical-dilemmas/

57. IBM. (n.d.). What is AI ethics? IBM. hhtps://ww.ibm.com/topics/ai-ethics

Chapter 11.

58. Ellingrud, K., Sanghvi, S., Dandona, G. S., Madgavkar, A., Chui, M., White, O., & Hasebe, P. (2023, July 26). Generative AI and the future of work in America. McKinsey Global Institute. https://www.mckinsey.com/mgi/our-research/generative-ai-and-the-future-of-work-in-america

59. Deloitte. (n.d.). Generative AI and the Future of Work. Deloitte. https://www2.deloitte.com/us/en/pages/consulting/articles/generative-ai-and-the-future-of-work.html

60. González Alarcón, N., & Melguizo, A. (2023, November 27). Skills for a Generative AI Future of Work: Insights and Strategies. UNESCO. https://www.unesco.org/en/articles/skills-generative-ai-future-work-insights-and-strategies

61. Tickner, A. (2024, February 26). Generation AI: New Skills for a New Era. Speexx. https://www.speexx.com/speexx-blog/generation-ai-new-skills-for-a-new-era/

62. Oliver Wyman Forum. (2024). How Generative AI Is Chang-

ing The Future Of Work. Oliver Wyman Forum. https://www.oliverwymanforum.com/global-consumer-sentiment/how-will-ai-affect-global-economics/workforce.html

Chapter 12.

63. Criticalcase. (2018, September 6). 7 Predictions on the Future of Artificial Intelligence and Machine Learning. Criticalcase. https://www.criticalcase.com/blog/artificial-intelligence-the-future-in-ten-years.html

64. Alavi, M., & Westerman, G. (2023, November 7). How Generative AI Will Transform Knowledge Work. Harvard Business Review. https://hbr.org/2023/11/how-generative-ai-will-transform-knowledge-work

65. Bergmann, D. (2024, February 9). The most important AI trends in 2024. IBM Blog. https://www.ibm.com/blog/artificial-intelligence-trends/

66. Forbes Technology Council. (2023, July 6). 20 New And Enhanced Roles AI Could Create. Forbes. https://www.forbes.com/sites/forbestechcouncil/2023/07/06/20-new-and-enhanced-roles-ai-could-create/

67. McKinsey Digital. (2023, June 14). The economic potential of generative AI: The next productivity frontier. McKinsey Digital. https://www.mckinsey.com/capabilities/mckinsey-digital/our-insights/the-economic-potential-of-generative-ai-the-next-productivity-frontier#business-value

68. Collaborations Between People and AI Systems (CPAIS). (2019, September). Human—AI Collaboration Framework and Case Studies. Partnership on AI. https://partnershiponai.org/wp-content/uploads/2021/08/CPAIS-Framework-and-Case-Studies-9-23.pdf

69. The HumanBots. (2023, September 12). Remote Work Revolution: How AI is Reshaping Business Operations. Linkedin. https://www.linkedin.com/pulse/remote-work-revolution-how-ai-reshaping-business-operations/

70. Mckinsey & Company. (2023, August 25). What is the future of Generative AI?. Mckinsey & Company. https://www.mckinsey.com/featured-insights/mckinsey-explainers/whats-the-future-of-generative-ai-an-early-view-in-15-chartsBonus: 100 ChatGPT Prompts to Get You Started

BONUS: 100 CHATGPT PROMPTS TO GET YOU STARTED

1. Navigating Policy Queries: Explaining Our Return/Exchange Process"How can I clearly articulate our return/exchange policy to clients based on this summary?" [Include policy summary]

2. Refining Communication: Enhancing an Executive Summary. "What improvements can be made to my executive summary for better clarity?" [Insert your executive summary]

3. Crafting Effective Meeting Resources: Sales Team Focus "I need assistance in developing a structured outline for a sales team meeting presentation." [Details about meeting objectives and attendees]

4. Innovative Team Bonding: Creative Activity Ideas "What unique team-building exercises could suit our department's size and interests?" [Information about the department]

5. Staying on Track: Project Proposal Deadline Reminder. "Can you remind me of the critical timeline for our project proposal submission?" [Project name and due date]

6. Market Insights: Exploring Digital Marketing Trends. "I'm looking for current trends in digital marketing. What should I focus on?" [Specify area or subtopic]

7. Event Planning Mastery: Organising the Annual Holiday Party. "I need help planning our company's holiday celebration." [Preferences or requirements for the party]

8. Content Creativity: Blogging on Customer Experience. "Could you propose some engaging blog topics related to customer experience?" [Specific focus areas or audience]

9. Data Organisation: Streamlining Customer Information. "How can I efficiently organise our customer data spreadsheet?" [Criteria or format details]

10. Learning the Ropes: CRM New Deal Creation Guide. "Could you provide a tutorial for adding a new deal in our CRM system?" [CRM details]

11. Celebrating Success: Employee Recognition Ideas. "What innovative methods can we use to appreciate our team's ef-

forts?" [Company culture and employee count]

12. Efficient Scheduling: Coordinating a Team Meeting. "Can you assist in finding a suitable time for a team meeting?" [Meeting duration, availability, preferences]

13. Team Synergy: Tools for Enhanced Collaboration. "What are some effective tools to boost our team's organisation and collaboration?" [Specific requirements]

14. Customer Support Excellence: Handling a Billing Issue. "I need help responding to a customer's inquiry regarding a billing matter." [Customer details and issue]

15. Inter-Departmental Communication: Strategies for Improvement"What are some dynamic ways to improve communication between departments?" [Current challenges]

16. Strategic Planning: Marketing Campaign Timeline. "Can you help me devise a timeline for our upcoming marketing initiative?" [Campaign dates, milestones, dependencies]

17. Email Campaigns: Enhancing Engagement. "I'm looking for tips to improve our email open and click-through rates." [Current challenges, metrics]

18. Sales Boost: Effective Lead Generation Methods. "What tactics could elevate our sales team's lead generation?" [Target audience or industry info]

19. Leadership Growth: Professional Development Resources. "Can you recommend resources for enhancing leadership skills?" [Focus areas or desired outcomes]

20. Digital Order: Structuring Our Shared Drive. "How can I best organise our shared digital files?" [Current categories or structure]

21. Marketing Creativity: Crafting Product Brochure Copy. "I need assistance in writing engaging content for our product brochure." [Key features or details]

22. Remote Team Unity: Virtual Team-Building Activities. "What are some effective online team-building activities for remote employees?" [Team size, available tools]

23. Project Efficiency: Software Recommendations. "Which project management software is ideal for improving our productivity?" [Features or integrations]

24. Competitive Analysis: Understanding Market Rivals. "Can you help us gather intel on our main competitors and their strategies?" [Industry, key competitors]

25. Wellness at Work: Employee Program Ideas. "What initiatives could we introduce for employee wellness and balance?" [Current programs or initiatives]

26. Customer Insights: Designing a Satisfaction Survey. "I need help creating a survey to gather customer feedback on our services." [Questions or metrics]

27. Customer Loyalty: Retention Strategy Suggestions. "What tactics can enhance our customer retention efforts?" [Churn rates, challenges]

28. Social Media Mastery: Scheduling Posts. "Can you assist in planning our social media content for the week?" [Platforms, content specifics]

29. Conversion Optimisation: Website Strategy. "What methods can boost our website's conversion rates?" [Focus areas]

30. Remote Collaboration: Effective Tool Selection. "Which tools can best facilitate remote team collaboration?" [Features or integrations]

31. Email Marketing: Crafting a Monthly Newsletter. "Can you guide me in designing our newsletter?" [Content or promotion details]

32. Large-Scale Bonding: Company Retreat Activities. "I need ideas for activities suitable for a large company retreat." [Location, theme]

33. Streamlining Operations: Simplifying Internal Processes. "What are some approaches to make our project approval process more efficient?" [Pain points, bottlenecks]

34. Advertising Success: Managing Google Ads. "Can you help optimise our Google Ads campaign for better results?" [Goals, audience]

35. Onboarding Enhancement: Customer Experience Improvement. "What can be done to better our customer onboarding process?" [Challenges, pain points]

36. Team Empowerment: Training Program Suggestions. "Could you recommend training to enhance our team's skills?" [Fo-

cus areas]

37. Insightful Analysis: Interpreting Sales Data. "I need assistance in analysing and summarising our sales figures." [Metrics, report specifics]

38. Sales Strategy: Upselling and Cross-Selling Ideas. "What are some effective cross-sell or upsell strategies for our products?" [Customer purchasing patterns]

39. Employee Engagement: Initiatives for a Motivated Workforce. "How can we boost employee engagement and morale?" [Survey feedback, challenges]

40. Targeted Marketing: Audience Analysis for Product Launch. "Can you assist in defining our target audience for a new product?" [Demographics, psychographics]

41. Time Management Mastery: Enhancing Productivity. "What are some effective time management strategies for better efficiency?" [Specific challenges]

42. Landing Page Efficacy: Lead Capture Optimisation. "How can I create a high-converting landing page for our campaign?" [Content, design preferences]

43. Team Recognition: Creative Reward Ideas. "What are some unique ways to acknowledge and reward our team's efforts?" [Budget, existing programs]

44. Customer Service Excellence: Best Practices. "Can you suggest strategies to improve our customer service and support?" [Feedback, improvement areas]

45. Social Media Strategy: Content Calendar Planning. "I need help with the next quarter's social media content plan." [Posting frequency, events]

46. Feedback Refinement: Enhancing Employee Reviews. "What are some methods to make our performance review process more impactful?" [Employee feedback]

47. Optimisation Tactics: A/B Testing Best Practices. "What are some effective approaches for A/B testing in our campaigns?" [Focus elements, metrics]

48. Customer Analysis: Segmenting for Targeted Marketing. "Could you guide me in segmenting our customer base for a marketing campaign?" [Segmentation criteria]

49. Engagement Strategies: Boosting Employee Involvement. "What initiatives could elevate employee engagement, especially in remote settings?" [Challenges, feedback]
50. Email Efficacy: Improving Deliverability Rates. "What are some tactics to enhance our email deliverability and avoid spam filters?" [Challenges, platforms]
51. Loyalty Building: Designing a Customer Rewards Program. "I need ideas for creating a customer loyalty program to encourage repeat business." [Rewards, incentives]
52. Website UX: Enhancing User Friendliness. "Can you suggest ways to improve our website's usability for a better user experience?" [Pain points, feedback]
53. Remote Communication: Tools for Team Connectivity. "Which tools can enhance communication among our remote workforce?" [Features, integrations]
54. Customer Insights: Crafting Effective Feedback Surveys. "How can I design a survey to gain valuable customer insights on our products?" [Questions, metrics]
55. Sales Elevation: Boosting Conversion Strategies. "What methods can we employ to improve our sales conversion rates?" [Sales processes, metrics]
56. SEO Strategies: Enhancing Online Visibility. "Can you recommend tactics to boost our website's SEO and attract more organic traffic?" [Keywords, competitors]
57. Referral Engagement: Creating a Customer Referral Program. "I need assistance developing a referral program to leverage our customer base." [Rewards, incentives]
58. Onboarding Optimisation: Employee Integration Improvement. "What can be done to make our employee onboarding process more effective?" [Challenges, pain points]
59. CRM Mastery: Customer Relationship Management Best Practices. "Can you suggest strategies for utilising our CRM system to nurture customer relationships?" [CRM specifics, goals]
60. Advocacy Activation: Building a Customer Advocacy Program. "How can I create a program to engage and mobilise our loyal customers for brand promotion?" [Rewards, customer base]

61. Learning Enhancement: Elevating Employee Training Programs. "What ideas can enhance our training programs for continuous employee development?" [Focus areas, initiatives]

62. Website Performance: Boosting Site Speed and Functionality. "What strategies can improve our website's performance and loading speed?" [Challenges, metrics]

63. Retention Focus: Crafting Customer Loyalty Campaigns. "I need help designing a campaign to reduce customer churn and foster loyalty." [Offers, messages]

64. Team Acknowledgment: Recognising Employee Contributions. "Can you propose creative ways to acknowledge our team's achievements?" [Team size, dynamics]

65. Collaboration Tools: Streamlining Remote Teamwork. "Which project management and collaboration tools are best for our remote team?" [Features, integrations]

66. Influencer Collaboration: Managing Influencer Marketing. "How can I identify and work with influencers to enhance our marketing campaigns?" [Audience, goals, budget]

67. Customer Journey: Enhancing Satisfaction and Experience. "What strategies can improve our overall customer experience and satisfaction?" [Pain points, feedback]

68. Internal Communication: Facilitating Better Team Interactions. "Can you suggest tools or strategies for improved team communication?" [Challenges, improvement areas]

69. Testimonial Leverage: Showcasing Customer Praise. "How can I effectively collect and utilise testimonials to highlight our product value?" [Channels, platforms]

70. Motivation Boost: Elevating Employee Morale. "What initiatives can enhance motivation and morale among our workforce?" [Challenges, feedback]

71. PPC Optimisation: Enhancing Campaign ROI. "Can you recommend tactics to improve the performance of our pay-per-click campaigns?" [Platforms, goals]

72. Social Media Content: Engaging Audience Strategies. "How can I curate and create content that resonates with our social media audience?" [Audience, themes]

73. Customer Self-Service: Enhancing Support Resources. "What

improvements can be made to our customer self-service resources, like FAQs?" [Pain points, feedback]

74. Workplace Diversity: Fostering Inclusion Initiatives. "What strategies can we implement to promote diversity and inclusion in our workplace?" [Challenges, improvement areas]

75. Customer Personalisation: Segmenting for Marketing Impact. "How can I segment our customer base for more personalised marketing campaigns?" [Segmentation criteria]

76. Feedback Evolution: Refining Employee Communication. "What ways can we improve our employee feedback and communication processes?" [Challenges, employee feedback]

77. Workflow Efficiency: Automating Business Processes. "What tools or technologies can streamline our business workflows for better efficiency?" [Pain points, inefficiencies]

78. Email Marketing Mastery: Optimising Automation Workflows. "Can you assist in setting up optimised email marketing workflows for better engagement?" [Segmentation, triggers]

79. Customer Retention: Loyalty Program Strategies. "What suggestions can you offer to improve our customer retention and loyalty programs?" [Challenges, program details]

80. Website Accessibility: Ensuring Inclusivity for All Users. "How can we make our website more accessible to users with disabilities?" [Accessibility guidelines, regulations]

81. Survey Design: Capturing Customer Feedback. "I need help designing a customer satisfaction survey for insightful feedback." [Questions, rating scales]

82. Team Collaboration: Fostering Effective Teamwork. "What are some methods to enhance employee collaboration and teamwork?" [Challenges, team dynamics]

83. Customer Journey Optimisation: Enhancing Touchpoints. "Can you suggest strategies to optimise our customer journey and improve touchpoints?" [Journey stages, touchpoints]

84. Rewards Program Creation: Incentivising Customer Loyalty. "How can I develop a rewards program to encourage repeat purchases and customer loyalty?" [Rewards, incentives]

85. Work-Life Harmony: Promoting Employee Wellness. "What

initiatives can we introduce to support work-life balance and employee wellness?" [Challenges, employee feedback]

86. Email Personalisation: Enhancing Marketing Automation. "Can you provide techniques for improving our email marketing automation and personalisation?" [Variables, segmentation]

87. Complaint Resolution: Designing an Effective Process. "How can I create an effective process for resolving customer complaints and feedback?" [Channels, procedures]

88. Conversion Funnel Optimisation: Improving Website Performance. "What are some strategies to optimise our website conversion funnels?" [Funnel stages, metrics]

89. Customer Service Efficiency: Reducing Response Times. "Can you recommend tools or strategies to improve our customer service response times?" [Challenges, response metrics]

90. Enhancing Team Communication: Overcoming Barriers. "What are some ways we can improve communication between different teams or departments?" [Challenges, improvement areas]

91. Productivity Boost: Project Management Tool Recommendations. "What project management tools can help streamline our workflow and improve productivity?" [Requirements, integration preferences]

92. Email Organisation: Managing Inbox Effectively. "Can you provide strategies for keeping my email inbox organised and managing high volume?" [Challenges, desired practices]

93. Meeting Efficiency: Crafting Structured Agendas. "How can I create an effective agenda for our weekly team meetings?" [Discussion topics, meeting goals]

94. Time Management Mastery: Navigating Busy Periods. "What tips do you have for managing time effectively during high workload periods?" [Specific challenges, constraints]

95. Technical Troubleshooting: CRM Connectivity Issues. "Can you assist with resolving a connection issue I'm experiencing with our CRM system?" [Error details, problem specifics]

96. Document Consistency: Creating Standard Templates. "How can I develop a standard template for our recurring sales

reports?" [Formatting, content requirements]

97. Effective Delegation: Empowering Team Members. "What are some best practices for delegating tasks effectively without micromanaging?" [Concerns, delegation challenges]

98. Customer Feedback Management: Organising Insights. "Can you guide me on how to categorise and prioritise customer feedback effectively?" [Feedback channels, categorisation criteria]

99. Performance Tracking: Setting Up KPI Dashboards. "I need assistance creating a dashboard to monitor and visualise key performance indicators for our sales team." [Specific KPIs, visualisation formats]

100. Content Strategy: Planning for Social Media. "Could you help me with a content strategy for our social media channels for the upcoming period?" [Target audience, key themes]

AI INTEGRATION STARTER TEMPLATE FOR NEWCOMERS

Welcome to the AI Integration Starter Template for Newcomers, a comprehensive guide designed to usher you into the world of Artificial Intelligence (AI) with clarity and confidence.

This template is designed to gently introduce AI newcomers to integrating AI into their professional lives, emphasising exploration and practical application as critical steps in the journey.

Instructions for Use:

- **Start Simple:**
  Focus on basic AI understanding before diving into more complex applications.
- **Be Curious:**
  Use your everyday tasks as a starting point for exploring how AI might make these easier or more efficient.
- **Seek Support:**
  Engage with communities and resources that cater to AI beginners. Be bold and ask questions, no matter how basic they may seem.
- **Iterate and Reflect:**
  Your AI integration journey is personal and evolving. Regularly revisit and adjust your learning goals and projects based on new insights.

Part 1: Introduction to AI

What is AI?

Write a brief description of what AI means to you after reading The Generative Edge.

____

____

____

Discover AI in Daily Life:

List examples of AI you've encountered daily (e.g., smartphone assistants, recommendations on streaming services).

Part 2: Understanding AI's Potential for Your Role

Identify Your Daily Tasks:

List out your regular job tasks or responsibilities.

Potential AI Application:

For each task identified above, utilise the table below to evaluate whether AI can improve, automate, or optimise the process. Should there be any uncertainty, indicate this with a question mark to denote areas requiring additional research. Subsequently, specify the AI tool that could facilitate the completion of the task at hand and describe the potential benefits such AI application could bring to your role in the workplace.

Workplace Task	Can AI Assist Yes/No	AI Tool	What are the Benefits/ Outcomes

The table below provides examples of various workplace tasks across different domains. It suggests associated generative AI tools that could support these tasks alongside the benefits or outcomes of using these tools.

Detailed Workplace Task	Generative AI Tool	Detailed Outcome/Benefits
Comprehensive Administrative Support (e.g., managing emails, organising schedules, handling routine queries)	ChatGPT for administrative tasks, x.ai for scheduling	Streamlines administrative processes by automating routine tasks, significantly reducing the time spent on emails and scheduling, leading to higher productivity, and allowing focus on strategic tasks.
Project Management and Coordination (e.g., delegating tasks, tracking project milestones, facilitating team collaboration)	AI-powered project management tools like Asana, integrating AI for task automation	Enhances overall project efficiency through automated task assignments and progress tracking, ensures timely project delivery by pre-emptively identifying bottlenecks, and improves team collaboration by providing actionable insights.
Advanced-Data Analysis and Interpretation (e.g., extracting actionable insights from large data sets, forecasting trends, and making data-driven decisions)	AI-driven analytics platforms like Tableau with AI capabilities	Facilitates informed decision-making by providing predictive analytics and trend forecasts, speeds up data processing and interpretation, and uncovers hidden insights that can lead to competitive advantages.
Creative Content Generation and Documentation (e.g., drafting reports, policy writing, generating marketing copy, designing visuals)	OpenAI's GPT for textual content creation, including reports and policy documents; DALL·E for image creation	Dramatically increases the output of high-quality creative and professional content, from detailed reports and comprehensive policy documents to engaging marketing materials and innovative visual designs, all while saving time and resources.

Detailed Workplace Task	**Generative AI Tool**	**Detailed Outcome/ Benefits**
Targeted Marketing and Campaign Optimization (e.g., enhancing customer segmentation, personalising marketing campaigns, optimising marketing strategies based on consumer behaviour analysis)	AI marketing platforms like HubSpot with AI for marketing automation	Enables more precise customer segmentation and personalised marketing campaigns, leading to higher engagement rates; optimises marketing strategies in real-time for better ROI; automates routine marketing tasks for efficiency.
Efficient Customer Service Management (e.g., providing instant support through chatbots, handling frequently asked questions, personalising customer interactions)	AI chatbots like Intercom with AI-powered responses	Offers round-the-clock customer support, ensuring immediate response to customer inquiries and issues; personalises customer interactions to enhance satisfaction and loyalty; frees up human agents for more complex queries.
Streamlined HR and Recruitment Processes (e.g., automating candidate screening, optimising interview scheduling, enhancing employee onboarding experiences)	AI recruitment tools like LinkedIn Recruiter with AI for matching candidates	Streamlines recruitment by efficiently screening candidates and matching them to job requirements; optimises interview scheduling to save time; enhances onboarding experiences through personalised learning paths.
Sales Enhancement and Customer Relationship Management (e.g., generating qualified leads, providing personalised customer experiences, predicting sales trends)	AI-enhanced CRM systems like Salesforce with Einstein AI	Improves lead generation and qualification processes, ensuring a higher conversion rate; provides personalised customer experiences to boost loyalty and retention; accurately predicts sales trends to inform strategy.

Learning About AI in Your Field:

Note in the table below any specific AI tools or technologies you've heard relevant to your industry. If none, make a plan to research (e.g., Google search "AI in [your industry]").

AI Tool/ Technology	Relevance to My Industry	My Plan to Research/ Explore

Use the above table to:

AI Tool/Technology:

List any AI tools or technologies you have heard of or are interested in.

Relevance to My Industry: Assess how each listed tool or technology is relevant to your industry.

My Plan to Research/Explore: Outline a research plan for each tool or technology. If you are unfamiliar with any, you can note down actions like "Google search 'AI in [my industry]'" or "Join a webinar on [specific AI tool]."

Part 3: AI Readiness for Beginners

Rate Your AI Knowledge:

Rate your current understanding of AI on a scale of 0 (no knowledge) to 10 (highly knowledgeable).

Part 4: Crafting Your AI Learning Path

Starting Small:

Pick one AI concept or tool you've heard about and would like to understand better. Write it down below.

Resource List:

For the concept/tool chosen, list at least two beginner-friendly resources to learn from outside the Generative Edge (e.g., YouTube tutorials and online articles).

Practical Exploration:

Plan a small, practical project or experiment you can do with this AI concept/tool, even if it's just using an AI writing assistant to draft emails.

Part 5: Identifying Where AI Fits in Your Role

Task Efficiency Review:

Revisit the list of your daily tasks and select one where AI's impact could be immediately beneficial. Describe how AI could assist in this task.

Implementation Steps:

Outline steps you might take to integrate AI into this task (e.g., using a specific AI tool or seeking assistance from a tech-savvy colleague).

ABOUT THE AUTHOR

Bailey Jackson is a working professional passionate about integrating cutting-edge technologies into everyday workflows to enhance productivity and efficiency. With extensive experience in leveraging Artificial Intelligence, particularly generative AI tools like ChatGPT, Bailey has transformed their own career and now aims to empower others to do the same.

The Generative Edge is a culmination of Bailey's journey and expertise, offering invaluable insights and practical strategies to help working professionals unlock unprecedented opportunities for growth and success. Bailey's approach is down-to-earth, relatable, and driven by a genuine desire to help others navigate and thrive in the rapidly evolving technological landscape.